The Institute o. _
Studies in Biology no. 10

Translocation in Plants

by Michael Richardson, B.Sc., Ph.D.

Lecturer in Botany, University of Durham

Edward Arnold (Publishers) Ltd

First published 1968
Reprinted with additions 1969
Reprinted 1971

Boards edition SBN: 7131 2188 2
Paper edition SBN: 7131 2189 0

Printed in Great Britain by
William Clowes and Sons Ltd
London, Beccles and Colchester

General Preface to the Series

It is no longer possible for one textbook to cover the whole field of Biology and to remain sufficiently up to date. At the same time students at school, and indeed those in their first year at universities, must be contemporary in their biological outlook and know where the most important developments are taking place.

The Biological Education Committee, set up jointly by the Royal Society and the Institute of Biology, is sponsoring, therefore, the production of a series of booklets dealing with limited biological topics in which recent progress has been most rapid and important.

A feature of the series is that the booklets indicate as clearly as possible the methods that have been employed in elucidating the problems with which they deal. There are suggestions for practical work for the student which should form a sound scientific basis for his understanding.

1968

INSTITUTE OF BIOLOGY
41 Queen's Gate
London, S.W.7.

Preface

The circulation of water, minerals and metabolites within plant tissues via the xylem and phloem was one of the earliest problems to attract the attention of plant physiologists. Studies on translocation, which have long been noted for the fascinating novelty and ingenuity of many of the techniques employed, received a great stimulus from the recent advent of radioactive isotopes, electron microscopy and methods involving the use of viruses and phloem-feeding aphids. Despite these recent advances, however, many problems and areas of dispute remain unresolved.

This booklet is designed to serve as an introduction to these problems, and to indicate how present-day research workers are attempting their solution.

Durham, 1968 M.R.

Contents

Introduction

The growing plant requires an extensive circulation of a wide variety of substances for the maintenance of its metabolic activities. In unicellular plants, such as some of the microscopic algae, the sites of food production and utilization are so close that translocation presents no problem. With the increasing specialization of multicellular plants, however, the photosynthetic tissues have in many cases become isolated in green aerial parts which are often situated at some considerable distance from the other tissues, such as the roots, which require the photosynthetic products. At the same time the green leaves require a constant supply of water and certain mineral elements which they cannot themselves acquire without the aid of the roots. To connect these separated tissues the plant has developed two distinct types of conducting elements which act as channels for the movement of substances. These are known as the xylem and the phloem.

In the past there has been a tendency to consider translocation in plants merely as an upward movement of water and absorbed mineral elements in the xylem and a downward movement of photosynthetic assimilates in the phloem. It is now known, however, that this type of approach to the problem is a naive oversimplification. Analysis of the contents of xylem sap has frequently revealed the presence of organic compounds (§ 2.3) and it is also known that a number of synthetic compounds such as antibiotics and growth regulators may be absorbed by plant roots and transported in the xylem (§ 2.5). Similarly, some inorganic ions may move in the phloem.

In addition to descending to the roots in the phloem, some of the sugars produced in the mature photosynthesizing leaves may also be moved upwards from these leaves to the growing apex and young leaves or developing reproductive tissues (flowers and fruits). It is now realized that this upward movement occurs in the phloem. The meristematic and actively growing parts of the plant may also import nitrogenous compounds from senescing tissues lower down the plant such as ageing leaves and petals.

Table 1 summarizes some of the complexities of the problem of translocation facing a plant. It is clear that many elements are translocated in a different chemical form from that in which they are absorbed or produced. For example, nitrogen is thought to enter plants from the soil mainly in the form of NO_3^- or NH_4^+ ions via the roots, but almost all of the nitrogen ascending the stem in the xylem does so in the form of organic molecules such as the amino acids, or amides. Similarly, at least some of the sulphur which enters the roots as the SO_4^{2-} ion is translocated in the form of the sulphur-containing amino acid, methionine, or the peptide, glutathione.

The picture is further complicated by the fact that mineral elements and plant products do not simply undergo translocation once in a single direc-

Table 1

Substance	Site of entry or production	Major functions	Form in which substance is translocated	Regions of plant in which substance is required
Water	Roots Stems and leaves (rarely)	1. Nutritive liquid or solvent 2. Turgidity 3. Synthesis of carbohydrates, etc.	As water	Everywhere
Mineral elements				
Nitrogen	Roots. Rarely leaves (Absorbed by roots as NO_3^- or NH_4^+)	Protein synthesis Nucleic acid synthesis Constituent of many other compounds (coenzymes, etc.)	Amino acids Amides. Other organic N compounds, rarely peptides and alkaloids Rarely as nitrate	Everywhere, particularly in meristematic tissues
Phosphorus	Roots (as PO_4^{3-})	Phospholipids Nucleic acids High-energy phosphate compounds Protein synthesis	Inorganic phosphate Phosphoryl choline } xylem Inorganic phosphate Sugar phosphates } phloem	Younger tissues Withdrawn from older, metabolically less active cells
Sulphur	Roots (as SO_4^{2-})	Proteins Constituent of vitamins (thiamine, biotin, CoA)	Inorganic sulphate Methionine and glutathione	Stem and root tips Young leaves Remobilized during senescence (protein breakdown)

Substance	Site of entry or production	Major functions	Form in which substance is translocated	Regions of plant in which substance is required
Iron	Roots (as Fe^{2+} or Fe^{3+})	Enzyme activity Cytochromes Chlorophyll synthesis	Possibly in a chelated form	Everywhere Collects along leaf veins
Calcium	Roots	Cell wall synthesis Enzyme activity Development of stem and root apices	Ca^{2+} ion (relatively immobile in plant) cannot move in phloem	Meristematic and differentiating tissues. Accumulates in older leaves
Magnesium	Roots	Photosynthesis (chlorophyll structure) Enzyme activity	Mg^{2+} ion	Leaves. Withdrawn from ageing leaves exported to developing seeds
Potassium	Roots	Maintenance of cell organization, permeability and hydration Enzyme activity	K^+ ion	Meristematic tissues Buds, leaves, root tips
Copper Zinc Molybdenum	Roots	Components of enzymes	Cu^{2+} Zn^{2+} ions Mo^{6+} (or Mo^{3+})	Everywhere Mo^{6+} particularly in roots
Manganese Boron	Roots	Activation of enzyme systems	Mn^{2+} ion B_3^{3-} or B_4O^{2-}	Leaves and seeds
Carbohydrates Sugars	Leaves and other green organs	1. Source of respiratory energy 2. Substrates for production of many plant constituents	Mostly as sucrose* Occasionally as raffinose, stachyose and verbascose Hexoses (glucose, fructose) not translocated	Roots. Meristematic and differentiating tissues. Flowers and fruits. Storage tissues

* See footnote on p. 7

Substance	Site of entry or production	Major functions	Form in which substance is translocated	Regions of plant in which substance is required
Growth regulators				
Auxin (indoleacetic acid)	Meristematic apices (root and shoot tips) Embryos, seeds, leaf buds, enlarging leaves	Stimulates elongation of stems and coleoptiles, cell enlargement. Regulation of differential growth rates (Tropisms and apical dominance)	As auxin. By active transport in a polar direction in non-vascular tissues of young, actively growing tissues As auxin in phloem of fully differentiated stems and petioles	Everywhere
Gibberellins	Mainly in roots Some seeds	Regulation of dormancy, flowering and fruiting Shoot growth Stem elongation	As gibberellins in xylem sap Possibly also in phloem	Everywhere, particularly in stem internodes and floral apices
Kinins	Roots	Stimulation of cell division, leaf and bud growth. Protein and nucleic acid changes	As kinins in xylem sap	Leaves. Shoot apical meristem Leaf buds
Vitamins				
Thiamine (B₁)	Mature leaves	Coenzyme for decarboxylases and transketolases	As thiamine in phloem	Roots
Pyridoxine (B₆)	Mature leaves Roots (?)	Cofactor for transaminase and amino acid decarboxylase enzymes	As pyridoxine in phloem	Roots and stem apex
Nicotinic acid	Leaves Possibly in some roots	Precursor of nucleotide coenzymes NAD and NADP	Unknown	Everywhere
Biotin	Young leaves	Coenzyme for carboxylases (fat synthesis)	As biotin in phloem	Roots

tion. In many instances they are subject to recirculation and redistribution. When the original supply of mineral elements ascending the stem in the xylem is more than adequate for the needs of the leaves and other aerial parts, the excess minerals may be translocated down to the roots again. BIDDULPH and BIDDULPH (1959) demonstrated this circulation of minerals by dividing the root system of a corn plant into two parts and feeding one half of the system with a nutrient solution containing radioactive phosphorus (^{32}P) whilst the other half was placed in a nutrient solution containing no radioactivity (Fig. 1–1). The radioactive phosphorus ascended

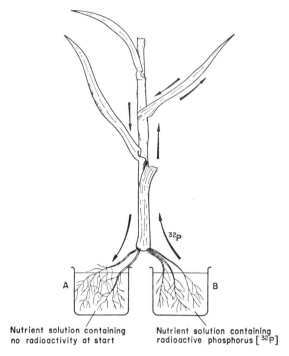

Nutrient solution containing   Nutrient solution containing
no radioactivity at start   radioactive phosphorus [^{32}P]

Fig. 1–1 The rapid circulation of radioactive phosphorus (^{32}P) within a plant. The isotope was absorbed from beaker B and transported upwards in the plant via the xylem. Within 6 hr radioactivity appeared in beaker A. (After BIDDULPH and BIDDULPH, 1959, *Scient. Am.*, **200**, Feb., 44–49.)

the stem, but within 6 hr it had descended the stem again and could be found in the solution which had previously contained no radioactivity. There is other good evidence which suggests that phosphorus undergoes a more or less continuous circulation in the plant via the xylem and phloem with only a small amount of the element being metabolically captured and incorporated into phosphorus compounds with low turn-over rates.

Successive radioautographs of plants which have previously been grown for a short time in a nutrient solution containing ^{32}P have shown that the isotope is rapidly translocated to all the tissues of the plant. Initially the tracer is most heavily deposited in the youngest leaves and buds, but as these tissues mature the phosphorus is withdrawn and retranslocated to other new and developing tissues.

Some other elements such as sulphur appear to make only one cycle within the plant body before complete metabolic capture occurs. At the other extreme, calcium seems to be immobile in the phloem and once delivered to a particular portion of the plant by the transpiration stream in the xylem, it remains there.

The need for a clear understanding of the translocation mechanisms within plants has a sound economic basis. There is much current interest in the foliar application to plants of herbicides, synthetic growth regulators and additional nutrients. Many of these new substances that are now commonly used in horticulture and agriculture depend for their effects on translocation. The transmission of many plant pathogens such as the fungi, bacteria and viruses and their reproductive spores within plants often involves the circulatory system. For example, recent work by PRESLEY et al. (1966) has confirmed that the wilt fungus *Verticillium albo-atrum* rapidly colonizes the tissues of the cotton plant as a result of its conidiospores being distributed by the transpiration stream in the xylem, rather than by mycelial growth. Certain other root-infecting fungi produce toxins which are distributed up the stem to the leaves by the transpiration stream in the xylem. Movement of many pathogenic viruses in plants involves both the xylem and phloem of the host. ESAU (1962) has described the complex and fascinating ways in which some viruses are introduced into the vascular tissues by insects (aphids, leaf-hoppers and white-flies) and subsequently moved with the food.

The two main problems of translocation which have attracted the attention of plant physiologists are firstly, the identity of the pathway of movement of specific substances, and secondly, the mechanisms or driving forces responsible for the movement.

The main reason why early studies on the pathways of movement gave so many conflicting results was the failure on the part of some investigators to recognize that the xylem, cambium and phloem tissues are in close spatial and physiological association. Lateral transfer of solutes from xylem to phloem, and vice versa, occurs readily, particularly in experiments of extended duration. In attempting to follow the pathways of transport of any compound in plants, precautions must be taken to minimize the lateral interchange between the xylem, cambium and phloem. Recent experiments with radioactive tracers where such precautions have been taken have greatly clarified modern ideas of the pathways of movement (§ 2.5 and § 4.3).

Whilst there is now general agreement as to the mechanism of movement of substances in the xylem (§ 3.2), there is still much fundamental conflict over the proposed mechanisms of movement in the phloem (§ 5.2–§ 5.6). A greater understanding of the structure of the phloem tissues, which is slowly being realized by the increasing use of the electron microscope, may eventually unravel the mystery of how they function.

Footnote to Table 1

* It is perhaps interesting to remark on the choice of sucrose as the major translocate of higher plants instead of the molecule glucose which plays such a central role in the circulation and metabolism of animals. ARNOLD (1968) has recently advanced the thesis that sucrose acts as a protected derivative of glucose. The ubiquitous distribution of enzymes affecting glucose would make this carbohydrate extremely vulnerable if it were a translocate. On the other hand sucrose is much less reactive. It is a non-reducing, easily hydrolysed derivative of glucose. Moreover the unusual translocates (raffinose, stachyose and verbascose) observed in some species are complex derivatives of glucose but include a sucrose molecule in their structure.

Pathway of Upward Movement of Water and Mineral Ions   2

2.1   Early ideas of xylem transport

The fact that the xylem is the tissue most involved in the upward movement of water and mineral ions has been recognized for well over a hundred years. As early as 1726 HALES drew attention to the movement of the sap within the xylem of plant stems and pointed out that the vessels and elements of the xylem were not restricted to the stem but continued via the small branches and petioles into the green leaves. At about the same time Magnol and De la Baisse supplied transpiring cut branches of plants with coloured fluids and thus mapped out the conducting tracts, which proved to be the vessels and elements of the xylem. By performing many experiments in which he measured the amount of water transpired by plants, HALES made the interesting discovery that 'the sunflower, bulk for bulk, imbibes and perspires seventeen times more fresh liquor than a man every 24 hours'. HALES and most of the early plant physiologists of the nineteenth century correctly concluded that this rapid flow or transpiration stream in the xylem provided a ready explanation for both the pathway and the mechanism of translocation of mineral ions from the soil to the aerial parts of the plant.

2.2   Structure of the xylem

It is a remarkable fact that the elements of the xylem which are the functional conduits for the movement of water and dissolved mineral ions are composed of dead cells whose central lumina contain no cytoplasm. This situation allows for the rapid and efficient translocation of large bulks of solution.

The main conducting or tracheary elements of the xylem are vessel elements and tracheids. At maturity when these cells are both dead and functional they are more or less elongated and possess lignified secondary walls. The individual vessel elements are distinguishable from tracheids by their shorter length and greater diameter (Fig. 2–1). In trees, vessels may range in diameter from about 20 μ to about 400 μ. In vines they may be as much as 700 μ in diameter.

Both vessel elements and tracheids have perforated end walls, but in the advanced vessel elements typical of the more highly evolved angiosperms the end walls may be entirely missing, thus leaving nothing to obstruct the passage of water through the cell. The side walls of vessel elements and tracheids are perforated by numerous pits. Generally the individual pits in

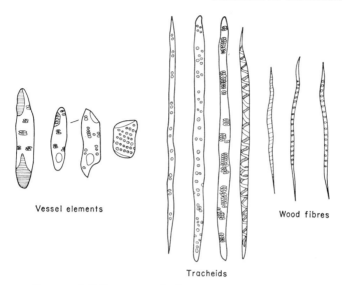

Vessel elements

Wood fibres

Tracheids

Fig. 2–1 Range of different vessel elements, tracheids and wood fibres found in xylem tissues. (After ESAU, 1953, *Plant Anatomy*. Wiley, New York and London.)

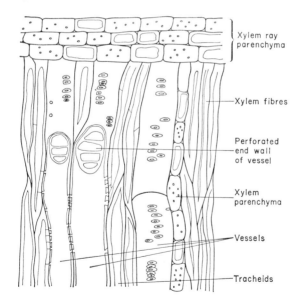

Xylem ray parenchyma

Xylem fibres

Perforated end wall of vessel

Xylem parenchyma

Vessels

Tracheids

Fig. 2–2 Radial section of a small portion of the xylem of an angiosperm. (After MEYER, ANDERSON and BOHNING, 1960, *Introduction to Plant Physiology*. Van Nostrand, Princeton, N. J. and London.)

one vessel element or tracheid are positioned in such a way that they are paired with the pits in adjacent vessel elements or tracheids, and are called accordingly pit pairs.

In angiosperms the vessel elements are arranged end to end with little or no overlapping and thus form a continuous pipe-like structure called the vessel (Fig. 2–2). Tracheids are also found in the wood of many but not all species of angiosperms.

The wood of gymnosperms is composed almost wholly of tracheids which occur in groups with much overlapping, and in such a way that the pits of adjacent cell walls are opposite each other (Fig. 2–3). Water can pass from cell to cell through these pits.

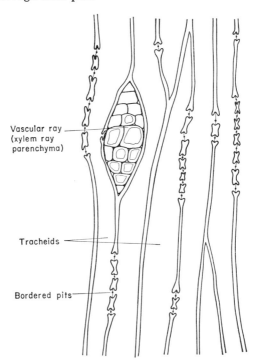

Vascular ray
(xylem ray
parenchyma)

Tracheids

Bordered pits

Fig. 2–3   Tangential section of a small section of the xylem of a gymnosperm.

In addition to the vessel elements and tracheids, the xylem tissues may also contain (wood) fibres and living xylem parenchyma cells. The xylem fibres tend to be thin, tapering cells which are generally longer than the accompanying tracheids (Fig. 2–1). The greater overlapping of these fibres together with their extensive and heavily lignified wall thickenings suggest that their primary function is one of support. It is doubtful if any significant amount of water is moved through these cells. The living xylem parenchyma cells are found either scattered between the vessel elements and tracheids (wood parenchyma) or as components of the xylem rays (ray parenchyma) (Fig. 2–2). The chief function of the xylem parenchyma cells appears to be the storage of food, as starch is built up in these cells during one growing season only to be utilized during the cambial activity of the following season. It seems unlikely that these living cells have any vital role in the translocation of water and mineral ions.

During the growth of many woody species some of the older xylem elements may cease to operate as channels for translocation. With increase in age numerous changes occur in the colour, composition and structure of the elements such that the sap-wood is converted to heart-wood. The water content of the heart-wood of many species is much reduced and the cells become characterized by the presence of oils, resins, gums and tannins. The relative importance of the sap-wood and heart-wood in the process of water movement has been investigated in a number of species of trees. Stems of oak (*Quercus*) trees have been rapidly frozen and then cut into thin transverse sections whilst still frozen. When this was done it was observed that although ice was present in most of the xylem elements, the clearest ice was seen in the vessels of the youngest growth ring. This result supported the view that the area of greatest water conduction is in the wood of the current year's growth. The water in the heart-wood of many species appears to be largely static and is not directly involved in translocation.

2.3　Composition of xylem sap

The most cogent evidence for the xylem as the main pathway or translocating tissue for mineral ions in the plant has resulted from analyses of the nature and concentration of the nutrients in the xylem sap.

There are several ways of obtaining xylem sap. Many plants grown under conditions favourable for water uptake but unfavourable for transpiration will exude liquid from points along the leaf margins. Small quantities of this guttation liquid can be collected and analysed, but the substances found in this fluid may not be a true indication of the contents of the xylem sap, as there is no direct continuity between the xylem elements and the guttation pores in the epidermis at the leaf margin. Better methods have been devised whereby sap may be collected from the exudations or 'bleedings' which occur when wounds or holes are made into the xylem of some species of trees. Unfortunately, only relatively few species of trees (e.g. *Acer saccharum*, *Betula*, *Vitis* and *Carpinus*) will bleed and even in these species the exudation is restricted to certain short seasons. On the other hand, exudates may be readily collected from the cut stumps of actively growing herbaceous species throughout most of the growing period. Xylem sap has also been obtained from a number of woody species by centrifugation of short lengths of wood, by displacement with a head of water, and by suction applied to a wound. Perhaps the most generally applicable method is the application of a vacuum to the lower end of a detached woody shoot, whilst small pieces are cut successively from the upper end of the shoot. Using this method it has been found possible to extract up to 25 per cent of the total moisture from the wood.

Analyses of the xylem saps obtained by these methods have generally revealed that they contain between 0·1 and 0·4 per cent solids of which approximately one-third is ash. Most of the essential mineral elements (N, P, K, S, Mg, Fe, Ca, Cu, Zn, Mo, Mn, B) can be detected at all times but marked seasonal variations tend to occur. Sugars and other organic constituents (amino acids, α-keto acids, alkaloids, coumarins) are frequently but not always present.

2.4 Ringing experiments

The earlier discovery that cut stems of plants would take up coloured solutions and dyes via the xylem, and the fact that the xylem sap itself contains mineral ions prompted plant physiologists to seek conclusive experimental evidence that the hollow cylinders of the xylem were the pathway of movement of inorganic ions.

At the beginning of this century various research workers supplied spectroscopically detectable cations such as lithium and caesium to the roots of plants and subsequently traced their movement by means of flame photometry. The upward transport of these elements was completely prevented by the removal of a short length of the xylem, whilst removal of a ring of phloem only partially inhibited the movement.

These ringing experiments clearly supported the view that the xylem was the normal channel for the upward movement of mineral ions, but widespread support for this idea was delayed by the apparent demonstration in 1935 by CURTIS that the upward movement of mineral salts might also occur in the phloem. CURTIS (1935) demonstrated that when a complete ring of bark was removed from branches of privet, the leaves above the ring made less growth and contained less ash and nitrogen than the leaves of control plants which were unringed. These results appeared to support the concept that the root-absorbed mineral salts were moving in the phloem, but later closer examination of the experimental technique employed by CURTIS suggested that the removal of a ring of phloem relatively high up on the stem probably interfered with the export of food substances from the leaves below the ring. Also the ringing may have partially starved the roots of their normal supply of carbohydrates from the leaves and this would tend to diminish the uptake of salts by the roots. Recent experiments have also shown that removal of a ring of phloem near the apex of the stem frequently reduces the rate of transpiration of the leaves above the ring and would thus interfere with the rate of supply of mineral ions if these were moving in the xylem sap.

2.5 The function of xylem

When radioactive elements became available for use as tracers STOUT and HOAGLAND (1939) provided very convincing evidence that the normal pathway of upward translocation of mineral ions from the soil is in the xylem. Small plants of cotton, geranium and willow, rooted in sand or solution culture were used. These authors carefully separated the phloem from the xylem on certain branches by delicately levering up 9 in. long rings of the phloem and inserting strips of impervious waxed paper between the xylem and phloem (Fig. 2–4). Great care was taken to avoid excessive damage to

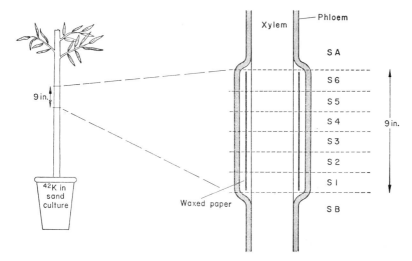

Fig. 2–4 Method of detecting upward channel of translocation of ^{42}K in a rooted cutting of willow (*Salix*). Lateral transport of isotope from xylem to phloem is prevented by a cylinder of waxed paper. Results may be seen in Table 2. (After STOUT and HOAGLAND, 1939, *Am. J. Bot.* **26**, 320–324.)

the tissues during the stripping operation, and the stripped areas were immediately covered with paper bindings to prevent undue loss of moisture. In certain branches the phloem was stripped from the xylem and then bound back into position without the intervening paper barrier. These branches were used as controls to show that the stripping operation had no apparent ill-effects on the transport up the stem. The plants were then allowed to absorb radioactive potassium (^{42}K) for a 5 hr period during which time conditions favourable for transpiration were induced by a fan.

After this period the branches were removed and the distribution of the tracer in the stem ascertained by measuring the amount of radioactivity in ashed segments of the xylem and phloem from the stem above, below, and in the region where the xylem and phloem were separated by waxed paper.

Table 2 Distribution of ^{42}K in the stem of willow (see also Fig. 2–4). (After STOUT and HOAGLAND, 1939, *Am. J. Bot.*, **26**, 320.)

		Stripped branch (xylem and phloem separated by waxed paper)		Intact branch	
		ppm ^{42}K in Phloem	ppm ^{42}K in Xylem	ppm ^{42}K in Phloem	ppm ^{42}K in Xylem
Above strip	SA	53	47	64	56
Stripped section	S6	11·6	119		
	S5	0·9	122		
	S4	0·7	112	87	69
	S3	0·3	98		
	S2	0·3	108		
	S1	20	113		
Below strip	SB	84	58	74	67

The results, shown in Table 2, clearly indicated that the radioactive potassium was translocated upwards in the xylem. However, the analyses of sections above (SA) and below (SB) the stripped area demonstrated that lateral transfer of the tracer from the xylem to the phloem could occur where these tissues were in contact with one another and not separated by a waxed paper barrier. The minute traces of radioactivity in the phloem separated from the xylem (S2–S5) suggest that translocation of ^{42}K is almost impossible in this tissue.

Examination of the results obtained for a control branch which remained unstripped with the xylem and phloem in contact, shows that slightly more radioactivity was found in the phloem than in the xylem, but we can now see that this was not due to translocation in the phloem but was almost entirely the result of a rapid lateral transfer of solutes from the xylem to the phloem.

In view of the fact that considerable lateral transfer of solutes occurred even in the relatively short time period (5 hr) of investigation employed by STOUT and HOAGLAND, it is perhaps easy to understand why so many conflicting results were obtained by earlier physiologists who made no attempt to separate the xylem from the phloem in experiments of extended duration.

The use of radioactive tracers has confirmed that most if not all mineral elements are transported upwards predominantly in the xylem. Certain complex organic molecules which are known to be produced or absorbed by roots are also thought to be translocated in the xylem.

SKOOG (1938) demonstrated that the natural growth regulator, indoleacetic acid (IAA) was readily absorbed through the intact roots of plants of tomato and squash. The upward movement of the compound was followed by measuring the degree of curvature which it induced at various heights up the stem (Fig. 2–5). The upward movement of the auxin was apparently in the xylem because the movement was unhindered when a section of the phloem was killed by steam. Similarly it seems reasonable to assume that

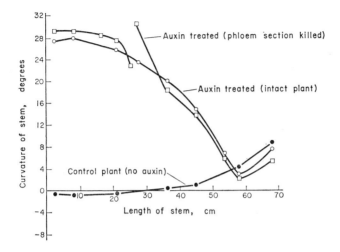

Fig. 2–5 Effect of killing a section of the phloem on the upward translocation of auxin (IAA) in squash plants. Curvature of stem is proportional to the amount of auxin translocated. (Adapted from SKOOG, 1938, *Am. J. Bot.*, **25**, 361–372.)

the translocation upwards from the roots of other growth regulators (naphthalene acetic acid, 2,4-D), antibiotics (streptomycin, penicillin, griseofulvin), systemic insecticides (Parathion, Demeton) and herbicides (maleic hydrazide, monuron) occurs in the xylem, as the upward movement of these compounds is little affected by phloem ringing. These compounds have frequently been detected in guttation drops and in the xylem exudates from excised roots. Moreover, the uptake and distribution through the plant of many of these compounds is varied directly by changes in transpiration rate (see § 3.1.)

2.6   Concept of carriers for mineral ions in xylem

It has generally been thought that the mineral elements which ascend the stem in the xylem do so in the inorganic form. However, there are a number of exceptions to this general rule.

Plants growing in soil usually absorb most of their nitrogen in the form of nitrate (NO_3^-), yet this compound can only be detected in the xylem sap of a very few species. Chromatographic examination of the nitrogenous constituents in xylem sap obtained from a wide range of plants has shown that the amino acids and amides (glutamine and asparagine) are always present, together with trace amounts of other nitrogenous compounds such as peptides and alkaloids.

Most of the sulphur absorbed by plants is translocated in the form of inorganic sulphate (SO_4^{2-}) but, in addition, some sulphur can be detected in the xylem sap as a constituent of the sulphur-containing amino acids, methionine and cysteine, or the tripeptide, glutathione.

Recent research has indicated that certain other ions may be translocated in the form of complexes with organic carrier molecules. TOLBERT and his co-workers (1955) have found that the bleeding xylem sap collected from species grown in culture solutions containing ^{32}P frequently contained (^{32}P) phosphoryl choline in addition to inorganic phosphate. Of the ^{32}P in the xylem sap some 6–20 per cent was present as this compound.

$$
\begin{array}{ccc}
CH_3 & & CH_2{-}CH_2 \\
 \diagdown & \diagup & \diagdown \\
CH_3{-}\overset{+}{N} & & O \\
 \diagup & \diagdown & \diagup \\
CH_3 & \overset{-}{O}{-}{-}{-}P & \\
 & O^{\diagup\diagup} \diagdown OH &
\end{array}
$$

The zwitterion structure and organic solubility of phosphoryl choline suggest that it might possibly act as a carrier in the xylem sap which was readily capable of penetrating cell membranes. As yet there is no evidence to support this idea, for although the addition of choline increased the phosphoryl choline content of the roots of plants growing in culture solutions, it decreased the overall transport of phosphorus to the shoot (TOLBERT and LOEWENBERG, 1958).

The movement of iron within the xylem may be restricted under certain circumstances in some plants. Plants grown in culture solutions containing a high ratio of phosphorus to iron, or at pH's greater that 7, frequently display symptoms of chlorosis which may be attributable to iron deficiency. Several workers have suggested that the immobility of iron under such conditions may be due to precipitation of the iron as ferric phosphate. It is known that the iron-deficiency chlorosis may be remedied by the application of iron to the culture solution as a chelated complex with ethylenediamine tetra-acetic acid (EDTA). This finding and the fact that plant species and varieties differ in their ability to absorb and translocate iron have led plant physiologists to suggest that some type of natural compound may function as a chelating agent and preserve the mobility of iron within the xylem.

Mechanism of Upward Movement in the Xylem   3

3.1 Influence of transpiration on xylem translocation

The mechanism responsible for the ascent of sap through the wood of trees has always fascinated botanists. One of the more intriguing questions was raised by the sheer size of some trees. Specimens of the American redwood (*Sequoia sempervirens*), the Douglas fir (*Pseudotsuga douglasii*) and the blue gum (*Eucalyptus*) frequently exceed 300 ft in height. Much of the experimental work on this problem has been performed on woody species since it was generally felt that a mechanism which could satisfactorily account for the movement of water and mineral salts in tall trees would be more than adequate to meet the needs of vascular species of lesser stature.

HALES (1769) was probably the first person to realize that there was a relationship between the transpiration of a plant and the upward movement of sap in the wood. Later workers observed that when an intact stem of a transpiring plant was cut beneath the surface of a dye solution, the dye rushed in, going both up and down the stem in the xylem. This inward rush of dye was thought to be due to the release of tensions or negative pressures within the elements of the xylem, which were presumably created

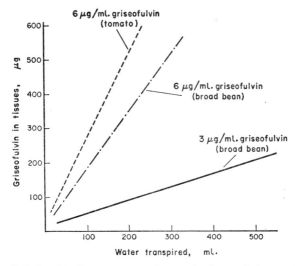

Fig. 3-1 Relationship between rate of transpiration and the uptake of the antibiotic griseofulvin by broad bean and tomato plants in water culture. (After CROWDY, GROVE, HEMMING and ROBINSON, 1956, *J. exp. Bot.*, **7**, 42–64.)

by the forces involved in transpiration since the dye failed to enter stems which had been defoliated. ZIMMERMAN (1963) has pointed out that these negative pressures within the xylem are probably also responsible for the faint hissing sound which may be heard when cuts are made into the xylem of many rapidly transpiring trees.

The upward movement of indicator substances such as spectroscopically detectable salts, dyes, and radioactive isotopes can usually be related to the rate of transpiration. Recently CROWDY and his co-workers (1956) have shown that the amount of the antibiotic griseofulvin taken up by broad bean and tomato plants in water culture was proportional to the volume of water transpired for any single concentration of the antibiotic (Fig. 3–1).

These observations are most readily explained by the cohesion–tension theory of sap ascent, which is the most plausible of the mechanisms proposed to account for the upward movement of water in plants.

3.2 The cohesion–tension theory of sap ascent

During transpiration water is lost from the leaves by evaporation from the micro-capillaries of the walls of the leaf parenchyma cells. This loss of moisture leads to the movement of water from the protoplasm and vacuole to the cell wall. This movement results in an increase in the osmotic concentration of the leaf parenchyma cell, which will in its turn osmotically attract water from adjoining cells of lower osmotic pressures. In this manner an osmotic gradient is built up across the leaf to the contents of the xylem elements of the leaf. The deficit of water in the terminal leaf parenchyma cell in this gradient will be satisfied by the withdrawal of water from the xylem element.

Molecules of water, although ceaselessly in motion, are also strongly attracted to each other. The existence of such inter-molecular attraction is particularly demonstrable when water is confined to relatively long tubes of narrow diameter, such as small capillaries or xylem vessel elements. If water is withdrawn from one end of such a system the water column is placed under tensile strain owing to the cohesive forces of attraction between the water molecules. These cohesive forces, together with the attraction which exists between the water molecules and the molecules of the wall of the tube (adhesive forces), prevent the water columns from breaking when they are subjected to a pull. Thus, as a consequence of the pull resulting from the loss of water from the leaves during transpiration, the water in the xylem ducts is drawn up the stem under a tension or negative pressure.

This view of the mechanism of sap ascent was first postulated by Dixon and Joly in 1894. Subsequently, DIXON (1914) demonstrated that the extracted sap of plants may have a cohesive or tensile strength considerably in excess of that of pure water. Sap collected from a branch of *Ilex aquifolium* by centrifugation was introduced into a J-shaped glass tube of about 1 cm diameter (Fig. 3–2). The tube was heated until the sap almost filled it and the small volume above the liquid contained only steam. At this point the tube was sealed. When the tube was gently inclined it was found possible to completely fill the long limb with a column of sap (Fig. 3–2b), which did not immediately rupture when the tube was again inverted (Fig. 3–2c). This failure of the column of sap to break immediately when inverted was due to the cohesive and adhesive forces of the water molecules. DIXON dis-

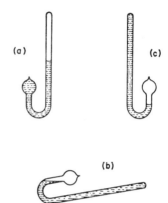

(a)

(c)

Fig. 3–2 Demonstration of tensile strength of xylem sap. (After DIXON, 1914, *Transpiration and the Ascent of Sap in Plants.* Macmillan, London.)

(b)

covered, however, that the column would rupture if the contents of the tube were cooled. From the degree of cooling required to break columns of several samples of sap, DIXON calculated that the tensile strength of sap from *Ilex* was in the range of 133–207 atmospheres.

Later workers have criticized the experimental technique which yielded these relatively high values for the cohesion of sap because the pressure in the tubes at the time of sealing was undoubtedly above the atmospheric level owing to the heating of the liquid. When later workers repeated this type of experiment under improved conditions where the pressure in the tubes was not above the atmospheric level at the time of sealing, they reported lower average values of 30 atmospheres for the tensile strength of water. Other workers who have employed biological methods rather than purely physical techniques have concluded that the water in fern annuli may support tensions of the order of 300 atmospheres prior to fracture.

The question remains whether these experimentally determined values are adequate to overcome the tensile forces to which columns of water in trees must be subject. Theoretically a cohesive force of 1 atmosphere should be sufficient to maintain a 32 ft (10 m) vertical column of water under ideal conditions. Thus the observed values should be more than adequate even for the tallest trees. For example, a 320-ft-high giant redwood would require approximately 10 atmospheres, a value which is well within the reported range. However, these theoretical calculations overlook the fact that water columns in trees are subject not only to vertical gravitational forces but also to the forces of resistance encountered when water attempts to move through narrow capillaries whose side walls are not smooth.

These additional forces of resistance may be illustrated by the fact that a 32 ft (10 m) log of a conifer lying in a horizontal position requires a pressure of approximately 1·5 atmospheres to drive water through it at a rate comparable with the normal rate of flow of the transpiration stream in the xylem. Likewise, similar logs of birch or maple in which the elements of the xylem conduits are short and irregular require pressures of the order of 3 atmospheres to overcome the forces of resistance in the xylem. Despite these additional forces, even the lowest values reported for the tensile strength of sap seem sufficient to maintain the continuous water columns in the xylem of most trees.

The cohesion–tension theory may be demonstrated by means of a simple physical system. A porous clay pot is filled with water and attached to the

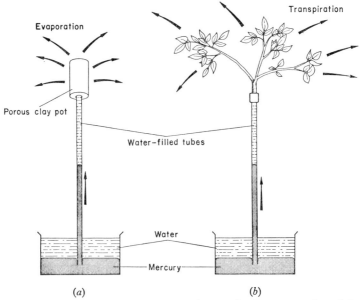

(a)

(b)

Fig. 3–3a Physical system for the demonstration of the cohesion–tension theory of sap ascent. Evaporation of water from the porous pot provides the tension required to pull up the column of mercury.

Fig. 3–3b Demonstration of the cohesion–tension theory of sap ascent. Transpiration of the leafy twig provides the tension necessary to pull up the column of mercury.

end of a long, narrow, glass tube also filled with water. The water-filled tube is placed with its lower end dipping below the surface of a volume of mercury contained in a beaker (Fig. 3–3a). A rotary fan may be used to move dry air over the surface of the porous pot. As water evaporates from the pores in the pot it is replaced by water which is pulled up through the narrow glass tube. Eventually the pull exerted in the continuous column of water will draw mercury up into the tube from the reservoir below. The cohesive forces between the water molecules, and the adhesive forces between the water molecules and the sides of the glass tube prevent fracture of the column. Using this type of model in 1893, Böhm demonstrated that the cohesive forces would support a 100 cm long column of mercury.

Dixon and Joly modified this experiment by replacing the porous pot with a twig cut from a pine tree. They noted that it was important to cut the twig from its parent plant beneath the surface of water to prevent the entry of air into the xylem elements. Provided this was done and the external conditions were favourable for rapid transpiration, the loss of water from the twig by transpiration was sufficient to draw up a 100 cm column of mercury (Fig. 3–3b).

We have other evidence that the contents of xylem vessels are in fact under tension in a normal transpiring plant. If the water in the xylem elements is under tension, it will, because of its adhesive properties, tend to cause a shrinkage in the diameter of these cells. Whilst this decrease in diameter is insignificant and cannot be measured for the individual xylem element, the total effect in the large number of xylem elements within the trunk of a tree can be measured by means of a dendrograph. In its simplest

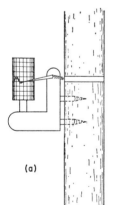

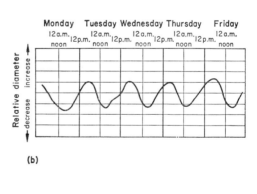

(a)   (b)

Fig. 3–4 (a) The dendrograph, a device for recording the diurnal variations in the diameter of trees. (b) Demonstration of diurnal variations in the diameter of the trunk of the Monterey pine (*Pinus radiata*). (a adapted from 'How Sap moves in Trees', by M. H. ZIMMERMANN, Copyright © 1963 by Scientific American, Inc. All rights reserved; b after MACDOUGAL, 1936, Carnegie Inst., Washington, Publ. No. 462, 55–60.)

form this instrument consists of a band or ring of metal fitted tightly around the circumference of a tree (Fig. 3–4a) in such a way that changes in the diameter of a tree may be recorded on a chart moving at a constant speed. Using such an instrument, MACDOUGAL (1936) recorded that the diameter of the trunk of the Monterey pine (*Pinus radiata*) was subject to diurnal periodic variations (Fig. 3–4b). During the daylight hours, when transpiration was at its greatest and hence the xylem elements were under their maximum tension, the trunk attained its minimum diameter. The trunk expanded again as the intensity of light decreased in the evening and the rate of transpiration fell. This phenomenon has since been shown to occur in the trunks of many other trees.

Another elegant demonstration that the sap is pulled upwards in the tree was developed by the German botanists, Huber and Schmidt, in 1936. They used an ingenious thermo-electric method whereby the contents of a xylem sap stream were briefly and moderately heated in one localized position and the upward movement of the heated sap determined by means of thermocouples placed at some distance from the point of application of

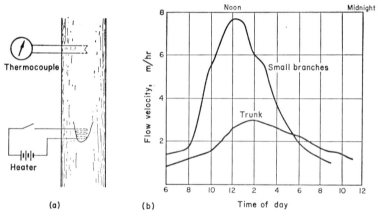

(a)   (b)   Time of day

Fig. 3–5 Demonstration that in the morning xylem sap flow increases in the small branches before the trunk. In the evening the flow rate decreases earlier in the branches than it does in the trunk. (Adapted from 'How Sap moves in Trees', by M. H. ZIMMERMANN. Copyright © 1963 by Scientific American, Inc. All rights reserved.)

the heat (Fig. 3–5a). They made the striking observation that in the morning water begins to move in the twigs and small branches high up on a tree before it does in the stem (Fig. 3–5b). On the other hand, in the late afternoon and evening as the rate of transpiration begins to slow down in the leaves, the sap movement decreases first of all in the twigs and only later in the stem. Logically therefore the driving force or 'motor' for the upward movement of water would seem to be located in the leaves at the top of the tree. This is in complete accord with the cohesion–tension theory.

It has frequently been pointed out that one of the more serious draw-backs of the cohesion–tension theory was that for its operation it required continuous columns of water, and the mechanism would cease to operate in the event of rupture of these columns by the formation of air bubbles. A single break will eliminate an entire vessel, a consideration of some significance when it is realized that these elements may extend over vertical distances of several feet. The xylem vessels of some trees undoubtedly contain air at certain times of the year. Bubbles of air might conceivably be introduced by the swaying and buffeting motion to which trees are subjected in high winds. It is also to be expected that bubbles of air may be forced out of solution when the contents of the xylem freeze during the winter. The air bubbles in these columns must be redissolved before they can become operational again. SCHOLANDER, LOVE and KANWISHER (1955) have suggested that the positive pressures which occur in the xylem during the spring must force the gas back into solution. In deciduous trees, such as the oak, even if the previous season's vessels have become blocked by air, they are supplemented by a new growth ring as a result of the activity of the cambium when the new leaves are formed.

A striking demonstration of the fact that trees can still maintain a rapid flow of sap even if many of their vessels become blocked by air was made by POSTLETHWAIT and ROGERS (1958). These workers injected a solution containing the radioactive isotope ^{32}P into the xylem of trees of pine and hickory (*Carya ovata*). Normally the isotope moved directly up the tree with little or no lateral displacement (Fig. 3–6a). However, prior to the

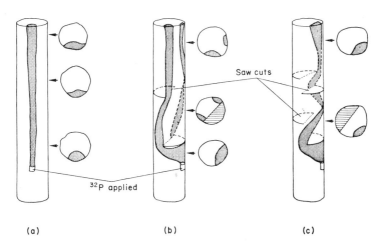

Saw cuts

^{32}P applied

(a)   (b)   (c)

Fig. 3–6   Movement of ^{32}P around saw cuts in the xylem of pine trees. Shaded areas indicate the location of the isotope. (a) No saw cuts; (b) two overlapping saw cuts; (c) four overlapping saw cuts. (After POSTLETHWAIT and ROGERS, 1958, *Am. J. Bot.*, **45**, 753–757.)

introduction of the isotope into the xylem of some of the trees deep saw cuts were made into the wood which was directly in the path of the ascending isotope. Under these circumstances the pathway of movement of the isotope avoided those elements into which air had been introduced by the saw cuts by moving laterally around the cuts in the most recently formed xylem (Fig. 3–6b and c).

In its basic essentials the cohesion–tension theory relies on a purely physical mechanism. Probably the only living cells which are directly involved are the cells of the transpiring leaf. KURTZMAN (1966) has recently shown that injection of trees with metabolic inhibitors and poisons such as picric acid and mercuric chloride fails to cause any significant change in the speed at which the sap flows.

3.3   The possible involvement of root pressure

Frequently when the stems are removed from trees or herbaceous plants the cut stumps which are left exude xylem sap. This phenomenon is said to be due to root pressure. The magnitude of this pressure may be demonstrated by attaching a manometer to the cut stump with a rubber sleeve (Fig. 3–7). Pressures of 3 to 5 atmospheres have been recorded by this method.

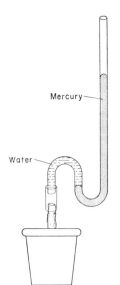

Mercury

Water

Fig. 3–7   Demonstration of root pressure in the cut stump of a plant.

Root pressure is referred to as an active process since it is dependent upon the metabolic activities of the root cells. Roots placed in an atmosphere deficient in oxygen have a considerably lower root pressure. Also, if roots are treated with potassium cyanide, which inhibits respiration, root pressure is again considerably reduced.

For a number of reasons it seems unlikely that root pressure plays any important part in the movement of moisture to the tops of tall trees. The recorded values of root pressure, which range from 3 to 5 atmospheres, are too small to account for the movement of water to the tops of most trees. Also, recent research has shown that the phenomenon is frequently absent in actively transpiring plants, and is invariably completely missing in conifers and other gymnosperms. Even in those herbaceous species in which root pressure has most often been demonstrated, the phenomenon only occurs when the excision is made in the lower reaches of the stem, relatively near to the roots. Finally, we have previously seen that the xylem sap under normal conditions is generally under a tension, whereas if root pressure was responsible for forcing water up the plant we would expect to find that contents of the xylem elements were under a positive pressure.

DIXON has suggested that during dormancy periods the process of root pressure may be responsible for the refilling of any vessels of the xylem which had become blocked with air during the previous season. However, this suggestion lacks experimental support and seems unlikely in most tall trees.

3.4   Regulation of translocation in the xylem

The dead cells of the xylem are separated from the phloem by a layer of living cells which is known as the vascular cambium. The primary function of a meristematic cell of the cambium is the continued production of new cells of xylem and phloem by successive divisions at a right angle to its radial axis. Another probable important function of the cambium, which is frequently overlooked, is that of control or regulation of the upward and lateral movement of solutes in the xylem.

As we have seen previously (§ 2.2), in many plants it is the most recently produced xylem lying adjacent to the cambium which carries the main stream of nutrient salts upwards. We have also noted the rapid lateral transfer of solutes which can occur between the xylem and the phloem (§ 2.5). This lateral movement must involve the intervening living cambial cells. It is also known that certain mineral nutrients are sometimes not delivered to the aerial parts of the plant in proportion to the volume of transpirational water which is delivered to them. This is not entirely unexpected as the different aerial tissues of the plant must have quite different qualitative and quantitative requirements. BIDDULPH (1959) has pointed out that an indiscriminative sweeping along of minerals in the transpiration stream would fail to accommodate the needs of certain tissues, whilst at the same time oversupplying other tissues. The living cells of the cambium are ideally situated to exert a metabolic and physical influence over the passage of minerals in the transpiration stream. It is assumed that this regulatory activity of the cambium is facilitated by the processes of active accumulation and secretion which require the expenditure of metabolic energy and thus only occur in living cells.

It is also feasible that some measure of control over the passage of solutes in the xylem may be exerted by the living cells of the wood and ray parenchyma.

Pathway of Movement of Organic Assimilates   4

4.1 Introduction

The discovery of blood circulation by Harvey in 1628 was a great stimulus to the study of translocation of organic solutes in plants. Prior to the mid-seventeenth century early plant physiologists thought that plants derived their nutrients entirely from the soil, but in 1679 Malpighi demonstrated that the 'ascending' or 'raw sap' from the soil was modified in the leaves through the action of sunlight, and thereby converted into an 'elaborated sap' which fed other parts of the plant. In his experiments Malpighi removed a complete ring of bark from the stems of a number of trees (Fig. 4–1a) and found that this girdling blocked the movement of the elaborated sap down the trunk, bringing about a swelling of the tissues above the ring (Fig. 4–1b). Malpighi also noticed that this type of swelling

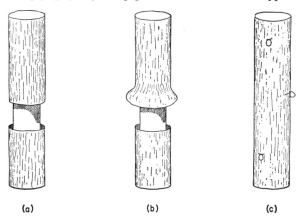

      (a)   (b)   (c)

Fig. 4–1 (a) Girdling of a woody shoot by removal of a ring of bark, leaving the xylem intact. (b) Swelling of bark a few weeks after the removal of a girdle. (c) Exudation of sap from cuts made into the bark of a tree.

due to the accumulation of the contents of the sap above the ring did not occur during the winter months when the trees lacked leaves.

Despite the evidence obtained from these ringing experiments it was still generally thought in the eighteenth century that the 'ascending' and 'descending' sap streams moved in the xylem. The significance of the phloem as the conducting tissue for the assimilates produced in the leaves escaped many early investigators. However, in 1837 Hartig discovered the sieve element in the secondary phloem of woody species, and in 1860 he observed that exudation of sap occurred when the bark of trees was cut into (Fig. 4–1c). Hartig immediately realized that the exudate was connected with the descending stream from the leaves and pointed out that it contained up to 33 per cent of sugar.

This important discovery of the role of the phloem as the pathway for the movement of organic assimilates failed to receive the immediate attention it deserved, because of the opposition it met from the eminent plant physiologist, Sachs. The then known structure of the phloem did not suit Sach's adherence to a diffusion theory of translocation whereas the structure of the xylem did. It was not until the 1930's that Hartig's ideas received due recognition as a result of new experiments with ringing, grafting, shading of foliage and removal of leaves (§ 4.3).

4.2   Phloem structure

The phloem tissue of most plants is composed of a number of morphologically quite different cells. The chief of these are the sieve elements, companion cells and phloem parenchyma, but in addition fibres, schlereids and albuminous cells also occur.

Investigation of the structure of the phloem has always been a relatively difficult task. Whereas the xylem tissues are composed mainly of dead cells with thick cell walls which are reasonably stable during the preparation of sections suitable for microscopy, the phloem is a delicate living tissue system whose cells are very easily damaged by even the most careful manipulation. Much of the uncertainty which has existed about phloem structure has stemmed from studies of mere artefacts produced by damage caused during sectioning and staining.

The widely accepted classical view of the structure of the 'mature' sieve element was that it was an empty or slime-filled tubular cell almost devoid of internal structure. Some recent studies of phloem structure by workers using either light or electron microscope techniques have indicated that mature sieve elements do appear to have a large cell lumen filled with slime which forms a continuous substance from cell to cell and little or no organized cytoplasm.

On the other hand, there is a considerable body of evidence accumulating from electron microscopy that the phloem of both angiosperms and gymnosperms during the initial stages of differentiation contains sieve elements whose cytoplasm possesses an abundance of well-ordered cell organelles, such as nuclei, vacuoles with normally developed tonoplasts, innumerable ribosomes, dictyosomes (golgi bodies), and endoplasmic reticulum (Fig. 4–2a and Plate 1).

As the sieve element undergoes enlargement and differentiation a number of marked changes occur in the structures of its cytoplasmic contents. The earliest recognizable sieve elements have their side and end walls

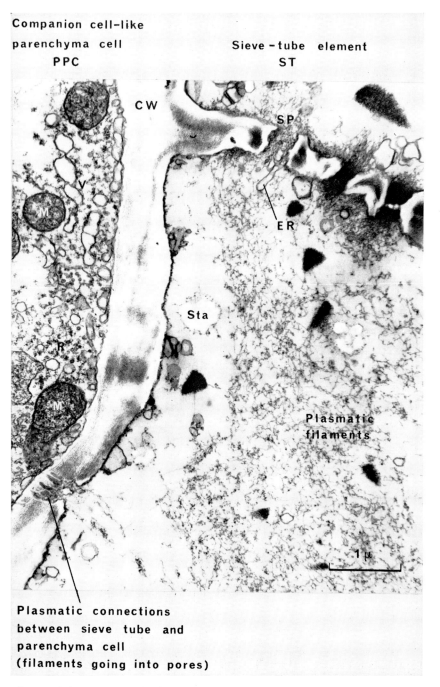

Companion cell-like
parenchyma cell
PPC

Sieve-tube element
ST

CW

SP

ER

Sta

Plasmatic
filaments

1 μ

Plasmatic connections
between sieve tube and
parenchyma cell
(filaments going into pores)

Plate 1 Ultrastructure of a sieve tube and an accompanying phloem parenchyma cell of *Dioscorea macroura*. **Key:** CW, cell wall; ER, tubular endoplasmic reticulum; M, mitochondria; PPC, phloem parenchyma cell; R, ribosomes; SP, sieve pore; ST, sieve tube; V, vesicular bodies. Magnification 20,000 ×. (Courtesy of DR. H-D. BEHNKE, Botanisches Institut, Universität Bonn.)

Plate 2 Aphid (*Tuberolachnus salignus*) feeding on the sieve tube contents of a stem of willow (*Salix*). (Courtesy of J. S. REDHEAD, Botany Dept., University of Durham.)

traversed by plasmodesmata (Fig. 4–2a). Groups of several plasmodesmata occur in each developing sieve pore and are surrounded by a small callose nodule. Tubules of endoplasmic reticulum which run throughout the cytoplasm also pass through these plasmodesmata. Sieve pores develop by the gradual removal of cell wall material in the region of the plasmodesma.

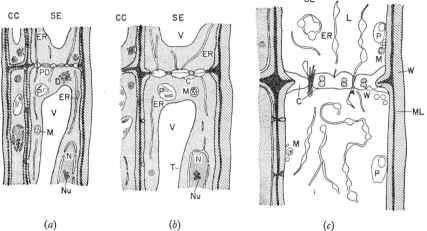

(a)   (b)   (c)

Fig. 4–2 Diagrammatic representation of structure of developmental stages of secondary phloem of *Pisum*, as interpreted from electron microscopy. (a) Earliest recognizable sieve element and companion cell; (b) intermediate stage of development; (c) morphologically fully expanded sieve element. **Key:** C, callose; CC, companion cell; D, golgi dictyosome; ER, endoplasmic reticulum; L, lumen; M, mitochondrion; ML, middle lamella; N, nucleus; Nu, nucleolus; P, plastid; PD, plasmodesmata; SE, sieve element; T, tonoplast; V, vacuole; W, cell wall. (After WARK and CHAMBERS, 1965, *Aust. J. Bot.*, **13**, 171–183.)

At the same time the callose deposits on the sieve plates increase and the cell walls become thicker (Fig. 4–2b). It is at this stage, when the sieve element is beginning to develop obvious sieve pores, that the nucleus begins to degenerate together with the dictyosomes and slime bodies. The internal structures of the endoplasmic reticulum, mitochondria and plastids may also become considerably altered.

During the next stages of differentiation the tonoplast begins to break down and the now less-dense cytoplasm containing modified mitochondria, plastids and vesicles becomes orientated along the long axis of the cell on the longitudinal walls (Fig. 4–2c). Coils of 'bead-like' endoplasmic reticulum remain scattered in the cell lumen and are occasionally observed to connect with the remnants of the plasmodesmata of the connecting strands still visible in the sieve pores. By this stage callose has been deposited over the cellulose walls between the sieve pores.

The most recent controversy to concern investigators of phloem structure has been brought about by the reported finding in a number of laboratories of discrete protoplasmic strands which crossed the lumina of 'anatomically mature' sieve elements and penetrated side and end walls via the sieve pores. THAINE (1962) has described these strands in mature sieve elements of *Cucurbita* and *Primula obconica* as 'transcellular strands' and indicated that they contain fibrils (which he suggests are possibly endoplasmic reticulum), mitochondria and plastids (Fig. 4–3). THAINE believes

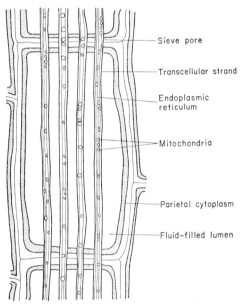

Sieve pore

Transcellular strand

Endoplasmic reticulum

Mitochondria

Parietal cytoplasm

Fluid-filled lumen

Fig. 4–3 Diagrammatic interpretation of structure of transcellular strands in a sieve tube element. (After THAINE, 1964, *J. exp. Bot.*, **15**, 470–484.)

that these transcellular strands are bounded by a tonoplast and that they completely fill the sieve pores between individual elements.

Several workers have been unable to find any evidence for such strands from studies using the electron microscope. Others who have described longitudinally straightened fibrils in electron-micrographs of sieve elements of *Pisum*, *Dioscorea* and *Metasequoia* believe them to be only elements of endoplasmic reticulum or longitudinal strands of slime which may or may not have a fibrillar structure, depending on the method of fixation employed. For example, TAMULEVITCH and EVERT (1966) reported the existence of strands in the phloem sieve elements of *Primula obconica* similar in appearance to those illustrated by THAINE. However, the strands were not membrane-bound cytoplasmic tubules; instead they consisted solely of slime and were derived ontogenetically from the slime bodies of immature cells. The individual units of slime possessed a tubular or fibrillar structure and were continuous from one cell to the next through the sieve pores but were never associated with mitochondria.

ESAU and her co-workers (1963) have also examined the structures in petioles of *Primula obconica* described by THAINE as transcellular strands. They think that the strands are merely artefacts probably due to lines caused by diffraction of light from cell walls out of focus. They support this contention by pointing out that the straight parallel lines or strands persist even in sections of cells whose cytoplasm was previously killed with sodium hydroxide or alcohol or removed by treatment with the proteolytic enzyme, papain.

However, THAINE maintains that, for a number of reasons, a distinction can be made between transcellular strands and diffraction lines. The transcellular strands do not have the uniform dimensions of diffraction lines. Linear files of moving particles may be seen in association with the strands and furthermore the particles inside the strands cause bulges when streaming stops. THAINE, PROBINE and DYER (1967) have also shown that when these strands are examined with an interference microscope they have the characteristic colour patterns exhibited only by real structures.

The contradictory results of these investigations and the varying opinions of the investigators are hardly surprising when one considers the number of variable factors involved: extreme sensitivity of the sieve-element protoplast to even minor disturbances, age differences of the sieve elements examined, the variety of species employed, and variation in methods of preparation and fixation.

4.3   Evidence for the movement of assimilates in the phloem

Hartig's discovery that the sap which exuded from cuts made into the phloem was rich in carbohydrates was confirmed in 1917 by Mangham who carried out micro-chemical investigations on sieve tubes and showed that the carbohydrate content of these cells varied in relation to the environmental conditions. Conditions which favoured rapid rates of photosynthesis in the leaves led to an increase in the carbohydrate content of the phloem sieve tubes.

A more careful analysis of the normal pathway for transport of soluble carbohydrates was made by MASON and MASKELL in 1928 using cotton (*Gossypium barbadense*) plants growing under field conditions. They removed a complete ring of phloem from plants in the morning and then analysed samples of both phloem and xylem from above and below the ring at intervals during the following 20 hr period. They found that ringing caused a marked accumulation of sugars in both the phloem and the xylem above the ring, and a marked decline in the phloem and the xylem below the ring (Fig. 4-4). Their results indicated that removal of the phloem certainly interfered with the downward movement of assimilates but did not, however, exclude the alternative hypothesis that the assimilates might have been moving in the xylem.

These early results were subject to a number of valid criticisms, the chief of which arose from the rather extended duration of the experiment. It is feasible that during the 20 hr period of the experiment removal of the phloem might have led to harmful effects on the movement of water and dissolved substances in the xylem.

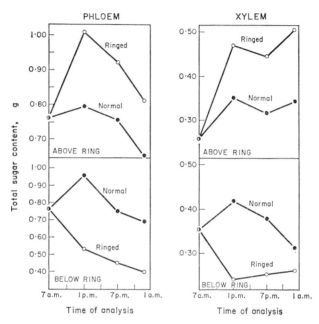

Fig. 4-4 Effect of removal of a ring of phloem on the concentration of sugars in the xylem and phloem (both above and below the ring) of cotton plants. (After MASON and MASKELL, 1928, *Ann. Bot.*, **42**, 190–253.)

MASON and MASKELL also showed that there was a diurnal variation in the sugar content of the sap in the various parts of the cotton plant (Fig. 4–5). Moreover, there was a much greater correlation between the varying contents observed in the leaves and in the phloem, than there was between the values for the leaves and the xylem.

In further experiments with cotton plants MASON and MASKELL levered up strips of bark and cut the phloem tissues away from the underlying xylem (Fig. 4–6). Some of the strips of phloem were bound back into place immediately without further treatment ('normal' plants), whilst others were first coated with a layer of Vaseline on their inner surfaces and then bound back into position ('Vaselined'). After varying periods of time the tissues of the bark and the wood of plants treated in these different ways were analysed for their contents of carbohydrates. In those plants where the xylem and phloem had been separated and then bound back into place

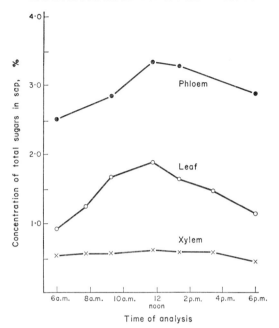

Fig. 4–5 Diurnal variation in the concentration of total sugars in the sap of the leaf, xylem and phloem of cotton plants. (After MASON and MASKELL, 1928, *Ann. Bot.*, **42**, 190–253.)

without an intervening layer of Vaseline, both the xylem and the phloem contained large quantities of carbohydrate ('normal plants', Fig. 4–6). Where the tissues were separated by a layer of Vaseline ('Vaselined') a much greater amount of carbohydrate was found in the phloem than in the xylem. Removal of a ring of phloem above the flapped section considerably reduced the carbohydrate content of the phloem in the flapped section. These results indicated strongly that the phloem was the normal channel of movement of carbohydrates in cotton plants. However, they also indicated that where the phloem and xylem were in intimate contact considerable lateral transport of carbohydrates occurred from the phloem into the xylem.

This lateral or 'sideways' movement of compound from the tissues in which they are normally transported into adjacent tissues in which they are not normally transported was probably one of the main reasons why many early workers were unable to decide clearly on the definite channels of transport for particular substances. Many of the investigators who followed MASON and MASKELL recognized the problem of lateral transfer and took steps to eliminate it during their experiments.

Just as the advent of the electron microscope was a tremendous stimulus to the study of the structure of the phloem, the use of radioactive isotopes

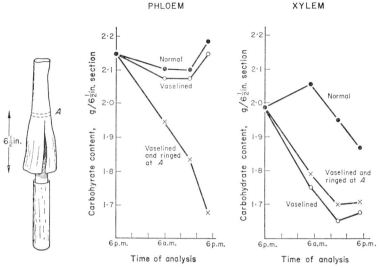

Fig. 4–6   Effect of separating the xylem and phloem by a layer of Vaseline on the translocation of carbohydrates in a cotton plant. Normal = section of phloem lifted and replaced without Vaseline; Vaselined = section lifted and replaced with xylem separated from phloem by a layer of Vaseline; Vaselined and ringed = same as 'Vaselined' but also ringed immediately above stripped section. (After MASON and MASKELL, 1928, *Ann. Bot.*, **42**, 190–253.)

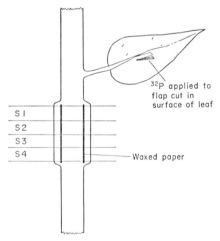

Fig. 4–7   Method of application of ^{32}P to leaves of cotton plants and arrangement of stripped sections where xylem and phloem were separated by waxed paper. For results, see Table 3. (After BIDDULPH and MARKLE, 1944, *Am. J. Bot.*, **31**, 65–71.)

as tracers greatly facilitated the convincing demonstrations by a number of physiologists that translocation downwards of organic assimilates occurred only in the phloem.

BIDDULPH and MARKLE (1944) studied the translocation of the radioactive isotope ^{32}P after it had been introduced into the leaves of cotton plants by immersing a small cut or flap on the surface of the leaf in the radioactive solution (Fig. 4–7). A slit was made into the bark of the stem below the leaf to which the ^{32}P was administered, and a piece of waxed paper was inserted in such a way that it completely separated the phloem tissues of the bark from the xylem (Fig. 4–7). After a short period of time the tissues which had been separated by the waxed paper were harvested and analysed for radioactivity. Only minute traces of radioactivity were found in the

Table 3   Distribution of ^{32}P between xylem and phloem of cotton plants during translocation from an injected leaf (see Fig. 4–7). (After BIDDULPH and MARKLE, 1944, *Am. J. Bot.*, **31**, 65.)

Section number	Per cent of migratory ^{32}PO$_4$			
	Stripped		Unstripped (control)	
	Phloem	Xylem	Phloem	Xylem
S1	12	1	15	5
S2	7	trace	10	6
S3	13	0	5	2
S4	5	trace	3	1

xylem, whilst much greater amounts were observed in the phloem (Table 3). These results clearly indicated that the downward movement of ^{32}P and ^{32}P-labelled organic compounds in plants occurred in the phloem. If, however, the phloem and xylem below the treated leaf were separated but then placed together again without inserting the waxed paper, then subsequent analysis revealed appreciable amounts of radioactivity in the xylem as well as the phloem (Table 3). Once again lateral transfer had occurred during the period of the experiment.

An improved method of introducing a radioactive tracer into a leaf, which did not involve the questionable technique of cutting into the leaf, was employed by RABIDEAU and BURR (1945). Single leaves of bean plants were allowed to photosynthesize in small glass chambers filled with an atmosphere containing ^{13}CO$_2$ (Fig. 4–8). Prior to the feeding with ^{13}CO$_2$ the phloem of some of the plants was killed by treatment of the stem above, below, and both above and below the leaf, with hot (100°) wax or a fine jet of steam. This effectively blocked the phloem. In those plants where the phloem was not blocked, the subsequent movement of the ^{13}C-labelled assimilates out of the leaf was found to be mainly up the stem to the young, actively grow-

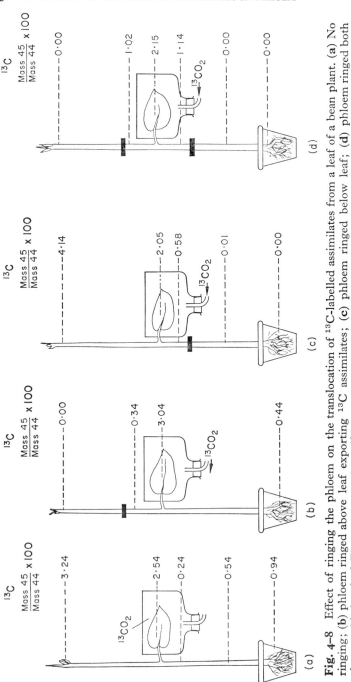

Fig. 4–8 Effect of ringing the phloem on the translocation of ^{13}C-labelled assimilates from a leaf of a bean plant. (**a**) No ringing; (**b**) phloem ringed above leaf exporting ^{13}C assimilates; (**c**) phloem ringed below leaf; (**d**) phloem ringed both above and below leaf. Figures given refer to ^{13}C content (Mass 45/Mass 44 × 100) of the tissues corrected for background. (After RABIDEAU and BURR, 1945, *Am. J. Bot.*, **32**, 349–356.)

ing regions (Fig. 4–8a) but some movement downwards towards the roots also occurred. When the phloem was blocked above the leaf, the movement of the isotope was predominantly down the stem (Fig. 4–8b); whilst blockage of the phloem below the leaf permitted only upward movement (Fig. 4–8c). If the plant had its phloem blocked both above and below the petiole, the ^{13}C assimilate did not penetrate past either of the rings (Fig. 4–8d), showing that it was being translocated in the phloem.

RABIDEAU and BURR demonstrated at the same time that ringing the phloem with hot wax or steam had not interfered with the movement of substances in the underlying xylem by simultaneously applying radioactive ^{32}P to the roots of the plant. The movement downwards of the ^{13}C assimilates was prevented by killing a ring of phloem, but the transport of ^{32}P up to the plant was only very slightly lowered by the existence of a ring of blocked phloem (Fig. 4–9).

Whilst it was conclusively demonstrated by these methods that the

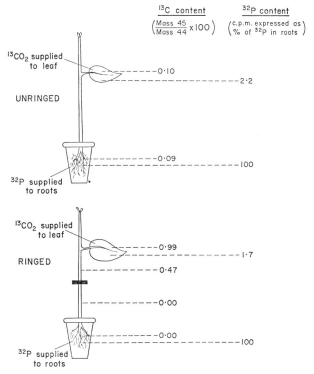

Fig. 4–9 Demonstration that removal of a ring of phloem prevents downward translocation of ^{13}C assimilate but does not interfere with the upward movement of ^{32}P from the roots. (After RABIDEAU and BURR, 1945, *Am. J. Bot.*, **32**, 349–356.)

phloem was the tissue in which the organic assimilates produced in the leaves were translocated both upwards and downwards, there was still some doubt as to which actual element of the phloem was involved. Early workers such as Hartig had hinted that it was the sieve tubes which were the actual channels of movement, and this was confirmed by BIDDULPH (1956) by the use of radioactive tracers and autoradiography of sections of the vascular tissues. BIDDULPH sprayed the leaves of red kidney bean plants with solutions containing the isotopes ^{32}P and ^{35}S and then after 1 hr made superimposed autoradiographs of transverse sections of the stems below the treated leaves. The autoradiographic plates showed areas of blackening which coincided with the phloem and in particular with the location of the sieve tubes in the sections. Similarly, in other experiments where leaves have been exposed to $^{14}CO_2$ micro-autoradiographic examination has shown that it is the sieve cells of conifers and the sieve tubes of angiosperms which are the channels of movement of the radioactive assimilate sucrose.

In recent years a number of different aphids and coccids, which are normally parasites on the translocation channels, have been used in experiments which clearly demonstrated the functional role of the sieve tube. The most widely used of these aphids is *Tuberolachnus salignus* (Plate 2), a relatively large aphid which colonizes stems of various species of willow (*Salix*). Other aphids which are currently being employed in research on translocation in both angiosperms and gymnosperms are *Cupressobium juniperi* (on juniper), *Cinara laricicola* (on Metasequoia), *Acyrthosiphon pisum* (on broad bean) and *Longistigma caryae* (on linden).

The aphid is allowed to insert its stylet tip into the phloem and then after 2 to 3 hr, by which time several droplets of honey-dew have been produced, the aphid is anaesthetized in a gentle stream of carbon dioxide. Whilst still anaesthetized the aphid is cut away from its inserted stylets with a fine splinter of glass. Generally sap exudes readily from these severed stylet stumps, which are left in the phloem, and may be collected with a fine capillary tube (Plate 3). Various experiments have shown that the exudates from such severed stylets are directly related to the materials being translocated in the phloem. Subsequent microscopic examination of sections of the stem on which the aphid was allowed to feed always revealed that it was the sieve tubes or a single file of sieve elements which were punctured and which supplied the exudate (Plate 4).

The aphid stylet method of determining the locality or region of functioning phloem cells and of obtaining uncontaminated samples of the contents is unsurpassed by any mechanical method as yet devised.

4.4 Translocation in lower plants

Our knowledge of the pathway and mechanisms of translocation of organic assimilates in the lower plants (algae, fungi, bryophytes) is relatively scanty. Nevertheless recent research has indicated that in some species at least there are close similarities to the pattern observed amongst the higher plants. For example, when ESCHRICH and STEINER (1967) applied bicarbonate—^{14}C to a single leaf of the moss *Polytrichum commune* L the radioactive tracer was incorporated into sucrose and rapidly translocated upwards. Autoradiographs of transverse sections of the moss showed that the radioactivity was confined to the leptoid cells of the stem. These cells quite closely resemble the sieve elements of higher plants in that they contain relatively dense protoplasts and their transverse end walls are covered by a thin layer of callose.

It has been known for some considerable time that cells resembling the sieve elements of higher plants occur in the stipes of the larger brown algae (Phaephyceae), and PARKER (1965) has recently shown that ^{14}C-labelled organic products of photosynthesis are translocated through these cells in the stipe of the giant kelp *Macrocystis* at rates of up to 65–78 cm/hr. However, the larger brown algae such as *Macrocystis*, *Nereocystis* and *Laminaria* also contain longitudinally directed medullary filaments which are sometimes known as 'trumpet hyphae'. The apparent morphological similarity of these cells to sieve tubes has led several research workers to suggest that they may also have a role in translocation. Although experimental data to support this contention is lacking, examination of the trumpet cells of *Laminaria* with the electron microscope has recently indicated a close homology with the sieve tubes of higher plants (ZIEGLER and RUCK, 1967). The cross-walls of the trumpet cells consist of a single primary pit field with 20,000–30,000 plasmodesmata whose pores are lined with a callose layer.

Unlike these larger brown algae, many smaller algae do not possess morphologically distinct tissues for translocation. Nevertheless, one of the most striking examples of the phenomenon of protoplasmic streaming may be observed in the stalk of the coenocytic green algae, *Acetabularia*. Examination of the stalk of this alga under the low power of the microscope reveals cytoplasmic streaming taking place in the form of many slender channels or striations running longitudinally inside the thin ectoplasmic layer. The flow in each striation is easily followed by the movement of chloroplasts which are carried along in trains. Most stalks contain 50–100 individual and parallel streams moving simultaneously in opposite directions in one and the same stalk. KAMIYA and KURODA (1966) have shown that the slender channels of flow may be detached from the striations as unit strands by centrifugation and have suggested that motile cytoplasmic fibrils may be the driving force behind the movement.

Whilst comparatively little is known about translocation in the fungi, another beautiful example of multistriate type of protoplasmic streaming as observed in *Acetabularia* occurs in the sporangiophores of the mucoraceous mould *Phycomyces blakesleeanus*. This and a variety of other different types of protoplasmic streaming in the fungi have been described by KAMIYA (1962).

Mechanisms of Movement in Phloem   5

5.1   Introduction

The problem of the mechanism of phloem transport is still the subject of appreciable controversy. Recent advances in our understanding of the structure of the phloem and the results of experiments with aphids, viruses and radioactive isotopes have tended to favour support of either the theories involving mass or pressure flow (§ 5.2) or the theory of protoplasmic streaming (§ 5.5). Nevertheless it would be premature on the basis of the existing evidence to emphasize one and to discard all the other hypotheses of phloem function.

5.2   The mass flow hypothesis

The hypothesis which views phloem transport of organic assimilates as a mass or pressure flow is probably the most widely supported of the four main ideas of the mechanism of movement in the plant phloem. The theory was initially postulated in 1860 by Hartig as an explanation of the exudation which he observed when he made cuts into the bark of trees, but it was more clearly formulated as a scientific theory by Münch in 1930.

In its original form the theory included all the living cells of the plant, but today its functioning is thought to be restricted to the sieve tubes.

During photosynthesis carbohydrates are produced in the leaves from carbon dioxide and water. In the translocatable form these carbohydrates exist as sucrose. Water which has ascended the stem in the xylem is absorbed by the leaf cells containing high concentrates of sucrose as a result of osmotic forces and this in turn brings about an increased hydrostatic (turgor) pressure in these cells. At the same time, a lowering of the concentration of sucrose in those regions where the assimilates are utilized for growth, storage and respiration results in a lowered hydrostatic pressure. Thus the regions of synthesis (green leaves) may be regarded as a 'source' and the regions of utilization as a 'sink'. Because of the gradients of hydrostatic pressure so created there will tend to be a bulk or 'mass' flow of solution and dissolved solutes from the 'source' to the 'sink', via the phloem.

Such a mechanism may be illustrated by means of a simple physical model (Fig. 5-1). The two cells, A and B, have membranes which are permeable only to water and are connected by a glass tube, C. Cell A contains a solution of high osmotic pressure, such as a solution of sucrose, and cell B contains only water. When the two cells are placed in water in a vessel, D, water will enter cell A from the external solution as a result of osmosis. This will create a hydrostatic or turgor pressure in cell A and cause solution to move out of the cell along the tube, C. This in turn will force water out of the second cell, B. The process will continue until the concentration of sucrose in both cells is equal, at which time the flow will cease. Cell A may be regarded as a 'source' and cell B as a 'sink'. If by

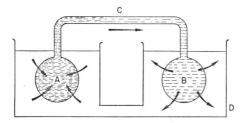

Fig. 5–1   Physical model to illustrate concept of mass flow in the phloem of plants.

some means one could replenish the sucrose in A, and at the same time continually remove it from B then the process would persist as a mass flow from A to B via the tube, C.

A mechanism involving mass flow might be one of the possible explanations of the phenomenon of phloem exudation, which occurs in many species when cuts are made into the bark. These exudates have been known since very early days, and are occasionally utilized as commercial sources of sugar. For example, a small industry in southern Italy is centred on the collection of the phloem exudates from trees of the ash family (*Fraxinus ornus* and *F. excelsior*). The bleeding of palm trees (*Arenga* and *Cocos*) has also been interpreted as a result of mass flow in the phloem.

The exudation which occurs from detached aphid mouth-parts inserted into the phloem frequently persists at a flow rate of about $1 \text{ mm}^3/\text{hr}$ for 2 to 3 days. This observation has been advanced as evidence for the occurrence of mass flow in the phloem. Experiments conducted by WEATHERLEY, PEEL and HILL (1959) showed that exudation observed from excised stylets of the aphid *Tuberolachnus salignus* on willow plants was the result of a longitudinal movement or flow of solution down the phloem sieve tubes. When cuts were made into the phloem directly above the point of insertion of the stylet the flow of exudate rapidly ceased (Fig. 5–2). However, if the cuts were made more than 15 cm above the stylet, or on either side of the point of attachment, then they had no effect on the exudation. These results clearly indicated that the aqueous solution exuding from the cut stylets was travelling a longitudinal distance of about 15 cm in the phloem, and was not derived from the cells immediately surrounding inserted mouth parts.

One of the basic requirements for the operation of the Münch theory of mass flow is that a gradient of turgor pressure must exist between the source and the sink. Such gradients of turgor pressure would be difficult to

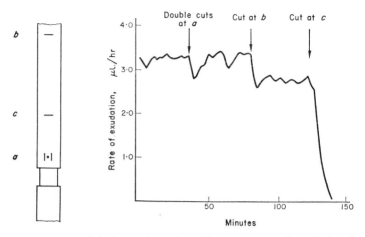

Fig. 5–2 Effect of incisions into the phloem on rate of exudation from excised stylets of the aphid *Tuberolachnus salignus* on willow plants. (After WEATHERLEY, PEEL and HILL, 1959, *J. exp. Bot.*, **10**, 1–16.)

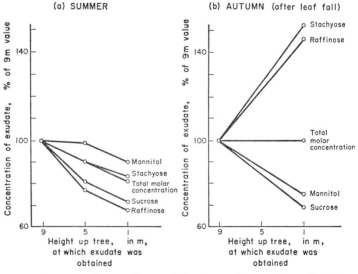

Fig. 5–3 Concentration gradients of the sieve tube exudates of white ash (*Fraxinus americana* L.). (a) Values obtained in the summer; (b) values obtained in the autumn (after abscission). (After ZIMMERMANN, 1957, *Plant Physiol.*, **32**, 399–404.)

demonstrate, but there is good evidence that concentration gradients do exist in the phloem tissues of various tree species. For example, ZIMMER-MANN (1957) has shown that definite gradients in the concentrations of the sugars sucrose, stachyose and raffinose, and the sugar alcohol, mannitol, exist along the phloem tissues of the white ash. Moreover these gradients were positive (decreasing concentration) in the downward direction of the trunk (Fig. 5–3a) during the summer months when the leaves were supplying photosynthetic assimilates. During the autumn and winter, however, after abscission of the leaves the concentration gradients of the individual sugars disappeared or became negative (Fig. 5–3b). Similar results were obtained when the trees were experimentally defoliated.

Under certain conditions a type of mass flow has been observed by several workers in microscopic sections of living sieve elements. BAUER (1953) immersed both of the cut ends of detached petioles of *Pelargonium* plants in solutions of varying concentration and subsequently observed surging movements of the contents of the sieve elements. Both the direction and the rate of movement were under the control of the artificially induced turgor gradients.

Perhaps the most convincing evidence in support of the mass flow theory has resulted from studies of the movement of extraneous substances in the phloem. It was consistently observed that when viruses or growth substances were applied to illuminated leaves they were rapidly translocated out of the leaves together with the assimilate stream in the phloem (BENNETT, 1937; ROHRBAUGH and RICE, 1949). However, when these extraneous substances were applied to shaded leaves then little or no translocation of virus or growth substance occurred (Table 4 and Fig. 5–4).

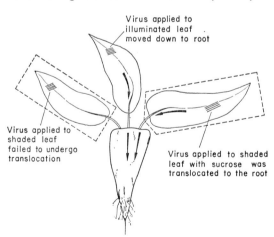

Virus applied to illuminated leaf moved down to root

Virus applied to shaded leaf failed to undergo translocation

Virus applied to shaded leaf with sucrose was translocated to the root

Fig. 5–4 Demonstration that translocation of curly-top virus from leaf to root in sugar beet requires the presence of photosynthetic assimilates in the leaf. (After BENNETT, 1937, *J. Agric. Res.*, **54**, 479–502.)

Table 4   Effect of light, dark and application of sucrose on the translocation of 2,4-dichlorophenoxyacetic acid (2,4-D) from trifoliate leaves of bean plants.   Stem curvature (°) indicates magnitude of translocation of 2,4-D. (After ROHRBAUGH and RICE, 1949, *Bot. Gaz.*, 111, 85.)

Treatment		No. of plants	Mean stem curvature (°)
In light	⎱ 70 μg 2,4-D	40	31·0
	⎰ 0 μg 2,4-D	20	0·0
Dark	⎱ 70 μg 2,4-D	40	0·0
	⎰ 0 μg 2,4-D	20	0·0
Dark plus 10% sucrose	⎱ 70 μg 2,4-D	40	20·4
	⎰ 0 μg 2,4-D	20	0·0

These observations strongly supported the idea that the sugars produced during photosynthesis in the leaves provided the turgor pressure gradients which were required for the operation of a mass flow. In the absence of photosynthesis the turgor pressure gradients soon disappeared and the virus or growth substances were unable to move out of the leaves. This idea received further support when it was shown that the viruses and growth substances could be translocated from the leaves in the dark provided that turgor pressure gradients were created artificially by the application of sugar solutions to the leaves (Table 4 and Fig. 5–4).

There is a possible alternative explanation of the requirement for sugars before translocation from the leaves can take place. If the process of translocation involves the expenditure of metabolic energy, then the energy required might be provided in the form of ATP as a result of the metabolic breakdown of the sugar during respiration. In the older theories of mass flow it was thought that a high pressure in the leaf mesophyll cells forced a solution of sucrose in water through the plasmodesmata and into the sieve tubes, but more recently it has been realized that this first stage of movement may actually occur against a concentration gradient (sugar concentration in the leaf mesophyll may be lower than the concentration in the upper sieve elements). If this is so the preliminary stage of mass flow requires an 'active transport' or secretion of sugars into the sieve elements. The energy for such a mechanism might be provided by the ATP produced from the respiration of the sugars. There is some evidence that this initial step does involve ATP or high energy phosphorylation, as a strong phosphatase enzyme activity has been detected in the leaf-vein sheath cells of the mesophyll tissues of many plants. Moreover ROHRBAUGH and RICE (1956) found that the translocation of ^{14}C-labelled 2,4-dichlorophenoxyacetic acid (2,4-D) from the leaves of tomato plants was severely reduced in plants which were deficient in phosphorus.

Plate 3 Collection of sieve tube sap exuding from an excised proboscis (containing stylets) of the aphid *Tuberolachnus salignus* feeding on willow (*Salix*). (Courtesy of J. FORD, Botany Dept, University of Hull.)

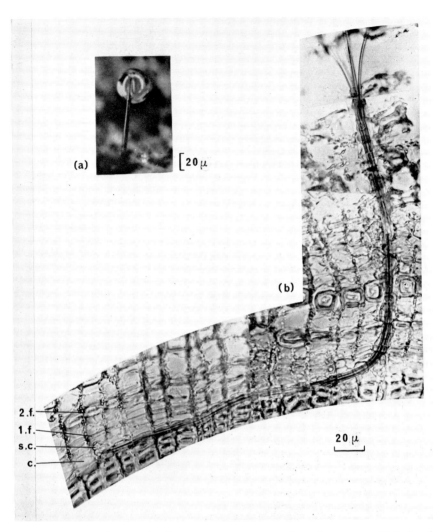

(a)

$20\,\mu$

(b)

2.f.
1.f.
s.c.
c.

$20\,\mu$

Plate 4 Photomicrograph of a transverse cross-section of a 2-year-old twig of *Juniperus communis* showing the inserted stylet of an aphid (*Cupressobium juniperi*) puncturing a phloem sieve cell. **Key:** f, fibre; sc, sieve cell; c, cambium. (From KOLLMANN and DORR, 1966, *Z. Pflanzenphysiologie*, **55**, 131–141; photograph courtesy of DR. R. KOLLMANN, Botanisches Institut, Universität Bonn.)

5.3 Electro-osmosis and mass flow

Two of the major objections to the Münch hypothesis of mass flow are (a) that it calls for excessive turgor pressure to account for the flow through the pores of the sieve plates, and (b) that it is essentially a non-physiological theory, whilst the phloem tissues themselves have a high physiological activity. Both of these objections have been answered by the electro-osmotic theory put forward by SPANNER (1958). In this theory two types of mass flow are envisaged: a mechanical mass flow down the lumen or vacuole of the sieve element due to turgor gradients, and an electrical or electro-osmotic mass flow through the pores of the sieve plates.

SPANNER postulates that an electro-osmotic flow of water accompanied by solutes through the sieve pores is brought about by polarization of the sieve plates due to uptake of an ion on one side and secretion of the same ion on the other side of the plate. SPANNER suggests that potassium (K^+) ions might fulfil such a role. The uptake of potassium on one side of the sieve plate and secretion on the other side could be effected by the adjacent phloem companion cells (Fig. 5–5). This unequal absorption and secretion

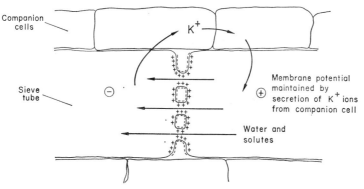

Fig. 5–5 Electro-osmotic flow of water and assimilates through the pores of phloem sieve plates. (After SPANNER, 1958, *J. exp. Bot.*, 9, 332–342.)

of a charged ion would create an electrical gradient which would produce a unidirectional flow through the sieve pores.

SPANNER feels that his electro-osmotic theory meets many of the major requirements demanded by the existing knowledge of the phloem structure and function. It is physically of the right scale, it is highly physiological, and it invokes well-known forces of adequate power for the task. It does not seem to involve any structural requirements which are not in fact known to exist and, conversely, at the time when it was proposed it assigned a role for many of the major features of the phloem which seemed to demand one.

Perhaps the best evidence for this mechanism is the recent demonstration by BOWLING (1968) that an electrical potential difference does exist across the sieve plates in the primary phloem of *Vitis vinifera*. This potential difference was measured by the insertion of micro-electrodes into the sieve tubes. The values observed, which ranged from 4–48 mV, were calculated to be of the right magnitude to sustain an electro-osmotic flow at the rates required.

Circulation of potassium ions around the sieve plate has been suggested as the mechanism which might cause the potential gradient. This ion is found in sieve tubes, and it occurs in concentrations which are approximately compatible with the suggested mechanism, but there appears to be some gross transport of potassium down the sieve tube. Nevertheless it is interesting to note that ROHRBAUGH and RICE (1958) have reported that the translocation of ^{14}C-labelled 2,4-D was greatly reduced in tomato plants starved of potassium prior to the experiment.

5·4   Objections to mass flow concept

Opponents of the mass flow hypothesis have raised a number of objections to its validity for all tissues. In its original form the theory demanded that the cells which supplied solutes to the flow (source) must have a higher turgor pressure than the cells which were receiving the flow (sink). Several workers observed that this was not always so. For example, it was seen that tissues such as cotyledons, ageing leaves and senescing petals which were relatively flaccid often exported foods to more turgid growing or storage tissues. Furthermore, Curtis and Schofield demonstrated that the osmotic concentrations of receiving tissues were frequently higher than those of supplying tissues. Present-day supporters of the mass flow theory have countered this type of objection by pointing out that the theory only requires a higher turgor pressure in the sieve elements in the supplying or source tissue and a lower turgor pressure in the sieve elements at the receiving or sink end of the channel of transport. Such conditions might be created by the rapid active transport of materials into the phloem at the source and removal of them from the phloem at the sink, and would be independent of the osmotic concentrations of the surrounding cells.

This type of mechanism has recently been demonstrated by BIELESKI (1966) who showed that excised phloem tissues of apple and willow could accumulate phosphate, sulphate and sucrose against a concentration gradient.

Münch's original theory assumed that the sieve tubes were passive or dead and that the sieve tube protoplasts played no active part in the transport mechanism. Since then, however, it has been shown that translocation is frequently associated with high levels of metabolic activity in the phloem. For example, in the absence of oxygen, translocation of sugars is invariably reduced or stopped altogether. Also the inhibitors of respiration such as potassium cyanide (KCN) and 2,4-dinitrophenol (DNP) inhibit transport in the phloem. Such findings are not necessarily incompatible with the revised ideas of the mass flow theory. Metabolic activity would be required for the process of active transport which is thought to maintain the concentration or turgor pressure gradient within the sieve elements. Furthermore, if some type of electro-osmotic force is operative at the sieve plates, metabolic activity might be required for the circulation of the charged ion about the sieve pores (Fig. 5–5).

A more serious objection to the mass flow theory resulted from the observation that solutes could move simultaneously both upwards and downwards in the phloem. A type of simultaneous bidirectional movement was demonstrated by PALMQUIST (1939) who supplied a dye fluorescein to a starch-depleted terminal leaflet of a compound bean leaf (Fig. 5–6).

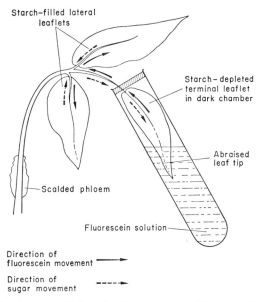

Direction of
fluorescein movement ━━━━━►

Direction of
sugar movement ━ ━ ━ ►

Fig. 5–6 Simultaneous bidirectional movement of fluorescein and sugars in the compound bean leaf. (After PALMQUIST, 1939, *Am. J. Bot.*, **26**, 665–667.)

Sections of the petiole taken at random times showed that the dye moved out of the terminal leaflet via the phloem whilst at the same time sufficient sugar moved out of the lateral leaflets and into the terminal leaflet for starch to be formed. If the outgoing dye and the incoming sugars were translocated simultaneously via the same phloem elements it would be very good evidence against the operation of a mass flow, but it is possible that the separate components of the bidirectional movement were occurring in different and independent sieve elements. The same provision applies to the demonstration by CHEN (1951) that different isotopically labelled compounds may move in opposite directions through the phloem at the same time. Figure 5–7 shows how a section of the phloem of a *Geranium* plant was separated from the xylem by the stripping technique of STOUT and HOAGLAND (see § 2.5), before $^{14}CO_2$ was administered to an upper leaf and $KH_2^{32}PO_4$ to a lower leaf. Analysis of this section after 15 hr revealed the

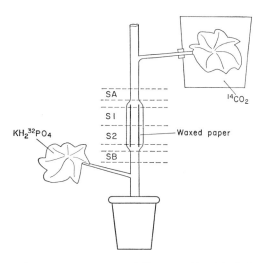

Fig. 5–7 Simultaneous movement of ^{32}P- and ^{14}C-labelled assimilates in opposite directions in the phloem of *Geranium*. See Table 5 for results. (After CHEN, 1951, *Am. J. Bot.*, **38**, 203–211.)

presence of both isotopes (Table 5) indicating that simultaneous bidirectional movement had probably occurred in the phloem during this experimental period.

Studies on the simultaneous movement of a number of different substances revealed that they do not all apparently move at the same rate. This type of evidence has also been used as an argument against the mass flow theory. For example, VERNON and ARONOFF (1952) supplied $^{14}CO_2$ to the leaves of soya bean plants and subsequently noted that the resulting radioactively labelled sucrose had a higher rate of translocation in the phloem than the radioactive glucose and fructose which were also produced. However, it is possible that the glucose and fructose which were observed were not being moved through the phloem but were merely the products of hydrolysis of the sucrose. BIDDULPH and CORY (1957) simultaneously applied water labelled with the isotope tritium (THO), ^{32}P and $^{14}CO_2$ to the surface of bean leaves and found that the apparent rates of translocation varied considerably for the different substances. These findings are only incompatible with the mass or bulk flow theory if there is no loss of the tracer compounds from the phloem tissues as they are transported. How-

Table 5 Simultaneous movement of ^{32}P- and ^{14}C-labelled assimilates in opposite directions in phloem of geranium plants, during a 15 hr period. (See Fig. 5–7 for location of sections.) Arrows indicate where ^{14}C and ^{32}P were supplied. (After CHEN, 1951, *Am. J. Bot.*, **38**, 203.)

Section of phloem analysed	Radioactivity	
	^{14}C (cpm/100 mg bark)	^{32}P (μg KH$_2^{32}$PO$_4$/100 mg bark)
	$\longrightarrow$	
SA (above waxed paper)	44,800	186
S1	3,480	103
S2	3,030	116
SB (below waxed paper)	2,380	125
	$\longrightarrow$	

ever, it is very likely that there will be unequal absorption of the different compounds from the phloem by the surrounding metabolically active cells. Supporters of the mass flow idea have drawn an analogy of phloem translocation with paper chromatography, where the solutes are moving as a bulk flow with the solvent in the paper but the different solutes undergo different degrees of absorption and retention by the paper.

5.5 Protoplasmic streaming in the phloem

A circular movement or 'cyclosis' of living protoplasm has been observed in many different plant cells. This protoplasmic streaming was first suggested as an explanation of the mechanism of phloem translocation by de Vries in 1885. The protoplasmic streaming theory has received much support from CURTIS and his associates who believe that the moving protoplasm carries the solutes within the sieve elements and that strands or layers of protoplasmic fluid may move from cell to cell through the comparatively large protoplasmic connections across the sieve plates (Fig. 5–8). Alternatively, the movement through the sieve pores may be due to the process of active transport.

Cytoplasmic streaming has never been observed in mature sieve elements, thus protoplasmic streaming as a possible translocation mechanism is limited to the younger phloem elements which contain metabolically active cytoplasm.

The best evidence for this theory would be the demonstration of simultaneous bidirectional movement within the smallest functional unit of conduction, i.e. within a single file of phloem sieve cells. The previously described demonstrations of bidirectional movement in the phloem

(PALMQUIST, 1939; CHEN, 1951) (see § 5.4) were not confined to single sieve elements.

BIDDULPH and CORY (1965) compared the export of ^{14}C-labelled assimilates from leaves in different positions on the stems of red kidney beans.

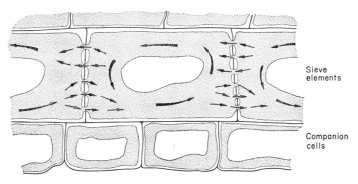

The lower leaves exported primarily to the root, the upper leaves to the stem apex, and the intermediate leaves in both directions. Bidirectional movement occurred in the younger phloem tissues, but was in separate phloem bundles. The upward-moving component was confined to bundles alternating with the downward-moving component from the next higher leaf.

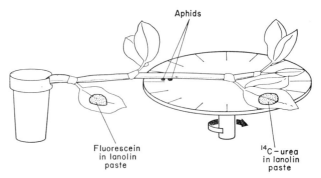

Fig. 5-9 Demonstration of simultaneous bidirectional translocation of fluorescein and ^{14}C urea in the phloem of *Vicia faba*. Honey-dew from the aphids (*Acyrthosiphon pisum*) was collected on a plate revolving at a rate of 1 rev./24 hr. (After ESCHRICH, 1967, *Planta*, **73**, 37–49.)

However, the possibility that simultaneous bidirectional movement may conceivably occur in a single sieve tube has recently been illustrated by ESCHRICH (1967). An upper leaf of a plant of *Vicia faba* was provided with a ^{14}C-labelled compound, whilst a lower leaf received the dye fluorescein. Figure 5-9 shows how the honey-dew was collected from aphids (*Acyrthosiphon pisum*) feeding on such a plant placed in a horizontal position over a revolving table. These aphids are known to puncture and feed only on the contents of a single sieve tube, yet the honey-dew produced in this experiment contained both tracers. At first sight this result seems to provide good evidence for simultaneous bidirectional movement in a single sieve tube and hence also for the operation of protoplasmic streaming as a mechanism. ESCHRICH has pointed out, however, the alternative possibility exists that the tracers were moving initially in opposite directions side by side in adjoining sieve tubes but subsequent lateral transfer of either tracer may have produced a loop path as illustrated by Fig. 5-10.

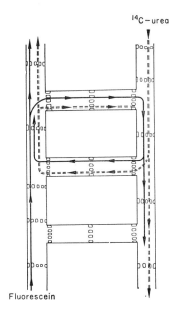

^{14}C-urea

Fluorescein

Fig. 5-10 Lateral transfer and loop path as an alternative explanation for apparent simultaneous bidirectional movement in a single sieve tube. (After ESCHRICH, 1967, *Planta*, **73**, 37-49.)

Advocates of the protoplasmic streaming hypothesis advance a great deal of evidence which suggests that translocation is an active process. For example, under certain conditions translocation may be blocked by respiratory inhibitors, such as KCN and azide, and narcotics, and diminished by lowered temperatures (2-3°C). Whilst these results do indicate that metabolic activity is required, they do not necessarily prove that protoplasmic streaming is involved. As we have already seen (§ 5.4), metabolic activity may be required for the active transport into and out of the sieve tube at the source and sink respectively, or alternatively the observed metabolism may be that of the phloem companion cells which are possibly connected with the maintenance of an electrical-potential gradient at the sieve pores (§ 5.3).

Recent experiments by FORD and PEEL (1966) have greatly extended our knowledge of the part played by metabolism in the movement of solutes within the sieve tubes. Certain portions of isolated segments of willow stems were cooled to a temperature of 2°C or less by means of chilled water or antifreeze circulating through an enclosing jacket, before ^{32}P-labelled phosphate was introduced into the phloem at one end of the segment. The movement of the isotope down the phloem was followed by measuring the amount of radioactivity which appeared in successive drops of honey-dew from an aphid sited below the cooled portion. It was found that the isotope continued to move down the stem even when the sieve tubes were cooled to $-1·5°C$, at which temperature it is known that protoplasmic streaming is completely prevented. It would seem, therefore, that the longitudinal movement of solutes within the sieve tube does not directly require metabolic activity. On the other hand, one noticeable effect of this lowered temperature was that it greatly reduced the movement of solutes out of the sieve tube across the lateral wall, thus indicating that the lateral transfer of solutes both into and out of the sieve tube may be dependent upon meta-bolic energy.

There are a number of reported observations which are difficult to re-concile with the protoplasmic streaming theory. Schumacher found that the dye fluorescein moved entirely independently of the streaming of cyto-plasm, and Arisz has reported that protoplasmic streaming in leaves of *Vallisneria spiralis* could be speeded up by wounding the leaves, but the increase in streaming did not increase the movement of Cl ions through the leaf. Similarly, when saponin was applied to coleoptile sections of *Avena* it stopped protoplasmic streaming without interfering with the transport of the auxin, indole-acetic acid (IAA).

The protoplasmic streaming theory of phloem transport has recently been extended to include a transcellular streaming concept. We have seen previously (§ 4.2) that THAINE (1962, 1964) believes that the lumina of phloem cells are commonly crossed by cytoplasmic (transcellular) strands which penetrate side and end walls. He has observed a type of protoplasmic streaming in fresh hand-cut sections of *Primula obconica* petiole phloem, where particles were seen in motion within the transcellular strands, passing longitudinally across the lumina and through sieve pores. Movement of the particles within a single strand was always in one direction, but the direc-tion was often both upwards and downwards in the same sieve tube element and even in neighbouring strands. THAINE suggests that the motive force or energy for this movement might be provided in the form of ATP by the mitochondria-like particles which he commonly observes within the trans-cellular strands (Fig. 4–3).

ESAU and her co-workers (1963) argue that the transcellular streaming filmed by THAINE does not occur within the sieve elements, but was probably due to cytoplasmic streaming in parenchyma cells, which were either above or directly beneath the sieve elements being examined. However, in 1967 THAINE and his co-workers provided further evidence in support of their view that translocation occurs within the transcellular strands. First, $^{14}CO_2$ was administered to a single leaf of a plant of *Cucurbita pepo*. They then collected the phloem exudate from a cut surface of the stem of this plant and subjected it to autoradiography. The phloem exudate examined in this manner contained straight-sided strands of radioactive material which were approximately $3 \cdot 6 \mu - 3 \cdot 8 \mu$ in diameter. The length of some of these radioactive strands (760μ) was approximately twice the length of a *Cucurbita* sieve element, which suggests that the exudate did not come from a single cell but may have been part of a strand of cytoplasm passing from one sieve element to the next through a sieve pore. The dimensions of the strands were, therefore, entirely consistent with the belief that they were the exuded contents of long, straight-sided, membranous structures in the phloem, such as the transcellular strands.

5.6   Other theories

As the result of a series of carefully conducted ringing experiments on cotton plants between 1928 and 1936, MASON and MASKELL concluded that the downward movement of sugar through the bark occurred along a concentration gradient. They suggested that the pattern of longitudinal movement was similar to that of diffusion, but occurred at a rate about 40,000 times faster than that of diffusion from a 2 per cent solution of sucrose. In 1937 MASON and PHILLIS proposed that the transport occurred through stationary cytoplasm by a process termed 'activated diffusion'. The living cytoplasm was thought to be capable in some way of hastening diffusion either by activating the diffusing molecule or by decreasing the resistance to diffusion through the protoplasmic medium. Unfortunately, such activation has not been demonstrated nor has any plausible explanation of the mechanism been put forward.

In 1932 Van den Honert proposed that certain solutes might be transported very rapidly along protoplasmic interfaces by diffusional gradients. It is a known fact that some substances can move extremely rapidly along the interface between two immiscible liquids. Van den Honert demonstrated this type of movement by layering ether over water containing a pH indicator and adjusted to pH $5 \cdot 8$ (Fig. 5–11). When a solution containing a small quantity of potassium oleate and excess potassium hydroxide was introduced at one end of the system it spread readily along the interface, producing an alkaline colour reaction as it proceeded. Using the colour change as an index of the rate of movement, Van den Honert observed velocities which were 68,000 times greater than those of simple diffusion.

He suggested that the vacuole–cytoplasm (water–lipid) interface of the phloem cells might be a pathway of rapid transport. Solutes such as sucrose adsorbed at the interface would lower the surface tension, and the motive power would be supplied by the removal of molecules at the end of

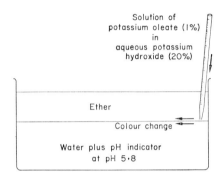

Fig. 5–11   Physical demonstration of surface (interface) theory of the mechanism of phloem translocation.

the interfacial path. This idea has received very little support as it is generally thought that the phase boundaries of typical phloem cells would provide insufficient surface area to account for the known quantities of solutes which are translocated in the phloem. Moreover, it is probable that too many different substances are involved in phloem translocation.

5.7   Rates of translocation

Measurements of the magnitude of translocation in the phloem are fraught with many difficulties. The most popular method has been to choose a plant organ such as a large developing fruit which, though green, increases in size and weight chiefly by virtue of the solutes translocated into it via the peduncle. One must bear in mind, however, that the increase in weight of such an organ might not only be a measure of the substances imported into it. The organ itself may additionally increase its weight by photosynthesis, or alternatively it may lose some weight as a result of respiration or export of foods from the organ. If the pathway of transport (phloem) in the connecting tissue (peduncle) has a uniform cross-sectional area then the mass transfer can be expressed as mass per unit cross-sectional area of the path. Table 6 shows the values for the mass transfer of dry weight that can be calculated from the published data on a number of different tissues by the equation:

$$\text{Rate of translocation} = \frac{\text{Transfer of dry weight per unit time}}{\text{Cross-sectional area of phloem}}$$

There is a striking similarity in the results for different stem tissues, and likewise for the petiole systems.

The values are calculated using only the cross-sectional area of the whole phloem tissue, but as we have seen previously (§ 4.3) there is good evidence that the sieve tube is the actual or main channel of movement within the phloem. Generally speaking the cross-sectional area of the sieve tubes occupies about one-fifth of the total cross-sectional area of the phloem, so the figures in Table 6 may be expressed as grammes per square centimetre

Table 6 Rates of translocation as measured by mass transfer of dry weight. (After CANNY, 1960, *Biol. Rev.*, **35**, 507.)

Plant system	Specific mass transfer g dry weight/cm² phloem/hr
A. Stems	
Potato tuber stem	2·1–4·5
Yam tuber stem	4·4
Cucurbita fruit peduncle	3·3–4·8
Kigelia fruit peduncle	2·6
B. Petioles	
Bean petiole	0·56–0·7
Tropaeolum petiole	0·7

of sieve tube per hour by multiplying them by five. When this is done the values for mass transfer of dry weight in stem give a mean value of 20 g/cm² sieve tube/hr.

Similar calculations can also be made for the mass transfer of sucrose in a single sieve tube, from the rates of exudation observed from excised aphid mouth-parts by WEATHERLEY, PEEL and HILL (1959). We know that the aphid mouth-parts only tap a single sieve tube (§ 4.3) They observed that approximately 1 μl. of a 10 per cent sucrose solution per hour was exuded from the mouth-parts of the aphid *Tuberolachnus salignus* on woody stems of willow whose sieve tubes had an average diameter of 23 μ. This represents 0·1 mg sucrose per hour per 414 μ² of sieve tubes or 24 g of sucrose per cm² of sieve tubes per hour. The value obtained in this perhaps more sophisticated way is in good agreement with the values presented in Table 6.

Many investigators have attempted to measure the velocity or rate of translocation of different substances in the phloem. This type of investigation is probably best attempted with radioactive tracers. The radioactive isotope is generally introduced into the plant via the leaf by injection or synthesis, and then the 'front' of the advancing radioactively labelled solute flow is detected lower down the stem. The time taken to travel a known measured distance is a measure of the velocity of translocation. However, there are a number of objections to this method. Firstly, allowance must be made for the time taken for the substance to enter the transport system, as an initial active uptake into the sieve cells may be involved (§ 5.4). Secondly, the method of detection of the 'front' must be highly sensitive to allow for the fact that the advancing front may not be very sharp or clearly defined.

An experiment of this type has been carried out by PEEL and WEATHERLEY (1962). Two aphid colonies were placed on the main axis of a willow cutting, one being placed just below the leaves, the other 30 cm or more further down the stem. Then $^{14}CO_2$ was fed to the leaves which were enclosed in a gas-tight glass chamber (Fig. 5–12). Samples of honey-dew

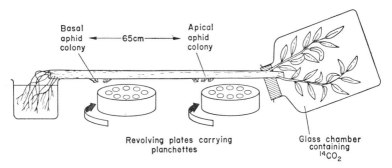

Basal aphid colony ←—65cm—→ Apical aphid colony

Revolving plates carrying planchettes

Glass chamber containing $^{14}CO_2$

Fig. 5–12   Method of estimating the rate of movement of ^{14}C assimilates in the phloem between two aphid colonies. (After PEEL and WEATHERLEY, 1962, *Ann. Bot.*, **26**, 633–646.)

from each of the colonies of aphids were collected directly onto planchettes at timed intervals and the radioactivities in the honey-dew determined by means of a Geiger-Muller end-window counter tube. Sampling was continued until the radioactivity of a sample from each colony was equal to at least twice the background radioactivity. This level was, of course, reached first by the upper colony, and the interval of time taken before the same level was attained by the lower colony was taken as a measure of the time necessary for the ^{14}C-labelled assimilate to travel the distance between the colonies. Table 7 shows the results of three such experiments. The values indicated that ^{14}C assimilates can move with a velocity of about 30 cm/hr in the sieve tubes of the willow (*Salix*). This result was only about a third of the rate previously indicated by the measured volume rate of an exuding stylet stump, but the discrepancy could be readily explained by exchange of the ^{14}C tracer with surrounding cells en route.

Analyses of data from experiments on many different species have revealed that in general the minimum velocities of translocation for normal assimilates in the phloem are of the order of magnitude of 10–100 cm/hr,

Table 7   Velocity of movement of ^{14}C assimilate between two aphid colonies. (After PEEL and WEATHERLEY, 1962, *Am. Bot.*, **26**, 633.)

Experiment number	Distance between aphid colonies (cm)	Time taken for radioactivity to travel from apical to basal colony (hr)	Rate of movement (cm/hr)
1	65	2·00	33
2	34	1·25	27
3	63	2·50	25

with occasional reports of values as high as 300 cm/hr for young seedlings. However, the wide variation of velocities often obtained within single experiments underlines the difficulties of measuring velocities, and points to the need for a large number of replicate experiments.

An example of this type of variation can be seen in the results obtained by Canny, Nairn and Harvey (1968) whilst investigating the velocity of translocation in 30 year old sycamore trees (*Acer pseudoplatanus* L.). These workers obtained clear evidence that the velocity of translocation varied considerably according to the season of the year during which the experiment was performed. The highest rates they observed were about 170 cm/hr for a short period in the autumn when the seeds were maturing. During most of the year the rates observed were only around 20 cm/hr.

Translocation between higher plants and their symbionts

Higher plants exhibit many forms of symbiotic association which may be either mutualistic, where both partners derive some benefit from the association, or parasitic in which case the association only benefits the parasite and is invariably detrimental in a greater or lesser degree to the host.

One of the primary features of a symbiotic association is the translocation of substances between the component organisms. Both the fungal parasites of the aerial parts of plants (such as the rusts, smuts and mildews) and the ectotrophic myccorhizal fungi derive their carbohydrates from the assimilates of the host plant. This has recently been demonstrated by means of ^{14}C autoradiography in a variety of symbiotic associations. The numerous experiments of this type have been summarised by SMITH, MUSCATINE and LEWIS (1969).

The fungal infection may markedly disrupt the normal pattern of translocation within the host plant. For example NELSON (1964) has shown that while the shoots of non-myccorhizal plants exported only 5% of their assimilates to the roots, those of myccorhizal plants exported 54%. Similarly LIVNE and DALY (1966) have demonstrated that bean leaves infected with the fungus *Uromyces phaseoli* exported only 2% of their photosynthate whereas uninfected controls exported over 50% during the same period. These alterations are presumably brought about by the creation of new 'sinks' in the vicinity of the fungal infection which maintain a concentration gradient between the host and the fungus.

Although sucrose appears to be the mobile carbohydrate which is extracted by the fungus, this host sugar is rapidly converted to fungal carbohydrates such as trehalose, mannitol, various other polyols and glycogen. There appears to be little or no reciprocal translocation of these metabolites from the fungus to the host.

There are a large number of angiosperms which are parasitic upon other higher plants. Some of these parasites completely lack chlorophyll and must obtain all of their carbohydrates from their hosts whilst others which do contain varying amounts of chlorophyll may require other essential substances from the host. This can be seen amongst members of the mistletoe family. In the case of the green mistletoes (*Viscum album* and *Phoradendron* spp) little or no detectable movement of photosynthates occurs between host and parasite, on the other hand in the dwarf mistletoes (*Arceuthobium* spp) considerable movement of assimilates has been observed. A similar picture has emerged from studies on the various members of the dodder family (*Cuscuta* spp).

There is no evidence that any of the chlorophyll-containing parasites translocate carbohydrate back to their hosts.

References

ARNOLD, W. N. (1968). *J. Theoret. Biol.* **21**, 13–20.
BAUER, L. (1953). *Planta*, **42**, 367–451.
BENNETT, C. W. (1937). *J. agric. Res.*, **54**, 479–502.
BIDDULPH, O. (1959). Translocation of inorganic solutes. In *Plant Physiology*, edited by F. C. STEWARD. Academic Press, New York and London.
BIDDULPH, O. and CORY, R. (1957). *Plant Physiol.*, **32**, 608–619.
BIDDULPH, O. and CORY, R. (1965). *Plant Physiol.*, **40**, 119–129.
BIDDULPH, O. and MARKLE, J. (1944). *Am. J. Bot.*, **31**, 65–71.
BIDDULPH, S. (1956). *Am. J. Bot.*, **43**, 143–148.
BIDDULPH, S. and BIDDULPH, O. (1959). *Scient. Am.*, **200**, February, 44–49.
BIELESKI, R. L. (1966). *Plant Physiol.*, **41**, 447–454.
BOWLING, D. J. F. (1968). *Planta*, **80**, 21–26.
CANNY, M. (1960). *Biol. Rev.*, **35**, 507–532.
CANNY, M. J., NAIRN, B. and HARVEY, M. (1968). *Aust. J. Bot.*, **16**, 479–485.
CHEN, S. L. (1951). *Am. J. Bot.*, **38**, 203–211.
CROWDY, S. H., GROVE, J. F., HEMMING, H. G. and ROBINSON, K. C. (1956). *J. exp. Bot.*, **7**, 42–64.
CURTIS, O. F. (1935). *The Translocation of Solutes in Plants.* McGraw-Hill, New York.
DIXON, H. H. (1914). *Transpiration and the Ascent of Sap in Plants.* Macmillan, London.
ESAU, K. (1962). *Plants, Viruses and Insects.* Oxford University Press, Oxford.
ESAU, K., ENGELMAN, E. M. and BISALPUTRA, T. (1963). *Planta*, **59**, 617–623.
ESCHRICH, W. (1967). *Planta*, **73**, 37–49.
ESCHRICH, W. and STEINER, M. (1967). *Planta*, **74**, 330–349.
FORD, J. and PEEL, A. J. (1966). *J. exp. Bot.*, **17**, 522–533.
HALES, S. (1769). *Vegetable Staticks*, 1726–1727, Vol. 1, 3rd ed. London.
KAMIYA, N. (1962). In *Handbuch der Pflanzenphysiologie*, XVII/2, edited by W. RUHLAND, Springer-Verlag, Berlin.
KAMIYA, N. and KURODA, K. (1966). *Bot. Mag.*, *Tokyo*, **79**, 706–713.
KURTZMANN, R. H. Jr. (1966). *Plant Physiol.*, **41**, 641–646.
LIVNE, A. and DALY, J. M. (1966). *Phytopathology* **56**, 170–175.
MacDOUGAL, D. T. (1936). Carnegie Inst., Washington, Publ. No. 462, 55–60.
MASON, T. G. and MASKELL, E. J. (1928). *Ann. Bot.*, **42**, 190–253.
NELSON, C. D. (1964). In *The Formation of Wood in Forest Trees*, edited by M. H. ZIMMERMANN, Academic Press, New York.
PALMQUIST, E. M. (1939). *Am. J. Bot.*, **26**, 665–667.
PARKER, B. C. (1965). *J. Phycol.*, **1**, 41–46.
PEEL, A. J. and WEATHERLEY, P. E. (1962). *Ann. Bot.*, **26**, 633–646.
POSTLETHWAIT, S. N. and ROGERS, B. (1958). *Am. J. Bot.*, **45**, 753–757.
PRESLEY, J. T., CARNS, H. R., TAYLOR, E. E. and SCHNATHORST, W. C. (1966). *Phytopathology*, **56**, 375–376.
RABIDEAU, G. S. and BURR, G. O. (1945). *Am. J. Bot.*, **32**, 349–356.
RICE, E. L. and ROHRBAUGH, L. M. (1958). *Plant Physiol.*, **33**, 300–303.

ROHRBAUGH, L. M. and RICE, E. L. (1949). *Bot. Gaz.*, **111**, 85–89.

ROHRBAUGH, L. M. and RICE, E. L. (1956). *Plant Physiol.*, **31**, 196–199.

SCHOLANDER, P. F., LOVE, W. E. and KANWISHER, J. W. (1955). *Plant Physiol.*, **30**, 93–104.

SKOOG, F. (1938). *Am. J. Bot.*, **25**, 361–372.

SMITH, D., MUSCATINE, L. and LEWIS, D. (1969). *Biol. Rev.* **44**, 17–90.

SPANNER, D. C. (1958). *J. exp. Bot.*, **9**, 332–342.

STOUT, P. R. and HOAGLAND, D. R. (1939). *Am. J. Bot.*, **26**, 320–324.

TAMULEVITCH, S. R. and EVERT, R. F. (1966). *Planta*, **69**, 319–337.

THAINE, R. (1962). *J. exp. Bot.*, **13**, 152–160.

THAINE, R. (1964). *J. exp. Bot.*, **15**, 470–484.

THAINE, R., PROBINE, M. C. and DYER, P. Y. (1967). *J. exp. Bot.*, **18**, 110–127.

TOLBERT, N. E. and LOEWENBERG, J. R. (1958). *Fedn Proc. Fedn Am. Socs exp. Biol.*, **17**, 323–331.

TOLBERT, N. E. and WIEBE, H. (1955). *Plant Physiol.*, **30**, 499–504.

VERNON, L. P. and ARONOFF, S. (1952). *Archs Biochem. Biophys.*, **36**, 383–398.

WEATHERLEY, P. E., PEEL, A. J. and HILL, G. P. (1959). *J. exp. Bot.*, **10**, 1–16.

ZIEGLER, H. and RUCK, I. (1967). *Planta*, **73**, 62–73.

ZIMMERMANN, M. H. (1957). *Plant Physiol.*, **32**, 399–404.

ZIMMERMANN, M. H. (1963). *Scient. Am.*, **208**, March, 132–142.